Rudolf Steiner

Mystery of the Universe: The Human Being, Model of Creation: Supplementary Materials

Edited by Frederick Amrine

Intermediate Anthroposophy 9

Contents

Introductions to Anthroposophy

Rudolf Steiner, "Excerpts from *The Boundaries of Science*." Trans. Frederick Amrine. ([Amazon:], Keryx, 2018).

Rudolf Steiner, *Anthroposophy and Mathematics*. Trans. Frederick Amrine. ([Amazon:], Keryx, 2019).

Basic Anthroposophy

Rudolf Steiner, CW 322; *The Boundaries of Science*. Trans. Frederick Amrine, foreword Saul Bellow, afterword Owen Barfield. ([Amazon:], Keryx, 2017).

Rudolf Steiner, GA 325; *Science and Historical Development since Antiquity*. ([Amazon:], Keryx, forthcoming).

Rudolf Steiner, GA 326; *The Origins of Natural Science*. Intro. Owen Barfield (Spring Valley, New York: Anthroposophic Press, 1985).

Intermediate Anthroposophy

Rudolf Steiner, GA 201; *Mystery of the Universe: The Human Being, Image of Creation* (London: Rudolf Steiner Press, 2001).

Frederick Amrine, GA 201; *Mystery of the Universe: The Human Being, Image of Creation: Supplemental Materials* ([Amazon:], Keryx, 2019).

Rudolf Steiner, CW 324a; *The Fourth Dimension: Sacred Geometry, Alchemy, and Mathematics*. Trans. Catherine E. Creeger (Great Barrington, Massachusetts: Anthroposophic Press, 2001).

Rudolf Steiner, GA 324; *Observation of Nature, Experiment, Mathematics and the Path of Anthroposophical Research*. Trans. Frederick Amrine. ([Amazon:], Keryx, forthcoming).

Advanced Anthroposophy

Rudolf Steiner, GA 320; *The Light Course: First Course on Natural Science*. Trans. Raoul Cansino (Great Barrington, Massachusetts, Anthroposophic Press, 2001).

Rudolf Steiner, GA 321; *Warmth Course*, 2nd edition (The Mercury Press, 1988).

Rudolf Steiner, GA 323; *Interdisciplinary Astronomy*. 3 vols. ([Amazon:] Keryx, 2019).

Frederick Amrine, GA 323; *Interdisciplinary Astronomy: Supplemental Materials* ([Amazon:] Keryx, 2019).

Frederick Amrine, GA 201; *Mystery of the Universe* and GA 323; *Interdisciplinary Astronomy: Commentary* ([Amazon:] Keryx, forthcoming).

Rudolf Steiner, CW 327; *Agriculture Course: The Birth of the Biodynamic Method* (London: Rudolf Steiner Press, 2004).

Introduction

These supplemental materials to GA 201; *Mystery of the Universe: The Human Being, Model of Creation* offer substantive emendations to the extant translation. They also supply color plates (produced in color however only in the electronic version), and notes, neither of which are contained in the RSP's version. In this important cycle of lectures, Steiner refers astronomy at every point to the human constitution. The result is an especially challenge, with rewards to match.

Perhaps the greatest hurdle to acceptance Steiner faced and still faces is that he occupies the seemingly excluded middle ground between science and religion. Steiner was himself a trained scientist, and he was deeply versed in both the history and the philosophy of science. But his own inner experiences confirmed to him the reality of the spiritual world, and he found there a rich field of phenomena that could be penetrated and understood by a researcher employing rigorous methods. The time has come, Steiner argued, when humanity must begin to transcend and supplant mere faith with, first, knowledge, and eventually direct experience, of spiritual realities. The mediation of this epochal transition is an important part of what Steiner meant by calling anthroposophy *spiritual science*. Partisans on both sides of this longstanding divide will inevitably be disappointed, but those who feel the pain of this deep wound in our humanity will gravitate towards anthroposophy.

Nothing about anthroposophy violates the spirit of modern science, which Steiner honors, but he rightly criticizes historical developments that arbitrarily restricted the ways in which science has come to understand itself and is practiced. Steiner reminds us that science is characterized (or should be) not by a predetermined set of permissible objects of inquiry, but rather by rigor, objectivity, and verification. The springs of modern science were clouded at their source by a desire to control nature, by unwarranted reductionism, and by the confusion of skepticism with rigor. Francis Bacon's triumphalist rhetoric would prove determinative: in his view the scientist should "omit no means of vexing" the goddess Natura, and "hound her in her wanderings." He imagines the scientist "leading to you Nature with all her children to bind her to your service and make her your slave." In the "Plan" of *The Great Instauration* (1620), Bacon boasts, "I do not propose merely to survey these regions in my mind, like an augur taking auspices, but to enter them like a general who means to take possession." Having treated nature as a slave and the spoil of war, is it any wonder that we are beset by ecological crises? Steiner foresaw them, and he offers solutions that go to the real root of the problem.

Following Descartes, modern science came to define its method arbitrarily as the elimination of all notions of causality but the mechanical, and to dismiss as "occult" phenomena that cannot be quantified. Hence, the Cartesians accused even Newton of having imported "occult qualities" into science, because his understanding of gravity and force implied no causal mechanisms. But the greatest scientists were not skeptics. Newton devoted as much time to Hermetic philosophy and theology as to math and physics. And Kepler (according to Kant, the most "rigorous" thinker who ever lived) claimed to have discovered, by his own intense efforts, the ancient wisdom that had been guarded and only partially revealed by Pythagoras and the Egyptian Mysteries.

Steiner worked out a scientific method for researching just those qualities, and his voluminous books, essays, and lectures report the results of his own noetic experiments. Steiner followed Goethe in recognizing that the perceiver is inextricably involved in the construction of experience; that all perception is already "theory-laden." For Goethe and Steiner both, the most precise scientific instrument is – the human being who has cultivated his or her faculties. Hence Goethe devised an alternative scientific method employing disciplined imagination, a rigorous science of qualities. The highest goals of science should be, not the disenchantment of nature to the end of controlling it, but rather the expanding of one's personal capacities in order gradually to enter into nature's wisdom. Like Goethe, Steiner felt that the ultimate goal of science should be *the transformation of the scientist*. As Freud, Husserl, and other of Steiner's contemporaries would also argue, there is no reason in principle why scientific rigor cannot be extended to the facts of our inner life. The ultimate goal of science is *theory* in the etymological sense: *theoria* comes from the same root as 'theater,' and it describes a contemplative viewing of spiritual facts. Self-transformation through meditative contemplation of phenomena is thus not the antithesis of science, but rather its ultimate goal and essence.

I plan to write a joint commentary on 201 and 323. This will be published eventually, but probably later rather than sooner.

Emendations to the Translation

Arabic numerals = page numbers

t = top; m = middle; b = bottom

boldface = suggested emendation

The original German title translates as *Correspondences between Microcosm and Macrocosm: The Human Being – A Hieroglyph of the Cosmos*.

Lecture 1

1m   There you have the natural scientific coloring of the **concept of necessity**.

2t   in which he applies himself to investigating **everything that belongs to the natural word order**, and the *Critique of **Practical** Reason*

3t   positive light upon matters with which **certain scientific disciplines** have to concern themselves.

4t   What has knowledge of the world become in the course of the last centuries **in the outer, popular world**?

4m   **Astronomers** reckon with the abstract three dimensions of space; they distinguish *one* dimension, a ***second*, and a** *third*, at right angles. **[Plate 1, left]**

4b   other than he finds in himself would be **a strong** illusion.

5m   conceive of an astronomical **paradigm** in the present abstract way.

5b   in a certain direction which adheres to the vertical. **[Plate 2, middle, above; Earth with a vertical mark]**

5b   as well as other movements, in a direction through which the earth also moves. **[The previous drawing is completed with the oscillating line.]**

6t   extremely difficult to train himself to feel that by **having something to do with these three dimensions of space** he is taking part in certain movements of the earth and the planetary system. **An anthroposophical** method of thought however can be **extended** to our knowledge of man by **seeking at least a substitute for this** understanding of the three dimensions.

6m   When we place a finger on **our** nose and move through this plane here **[Plate 2, right]**, we are moving through the vertical symmetry

6b   We see with two eyes, in such a way that the lines of vision intersect. **[Plate 2, left]**

6b   is divided off by a plane which stands at right angles to the first. **[Plate 2, 2nd vertical plane]**

7t   Thus we should obtain a third plane which again is at right angles to the other two, **that is horizontal and would pass through your arms when you hold them** *thus* **[stretched out to the side, the surface of the hands pointing downward]. Your hands would then fall within this plane.**

7m   **In the end,** everything through which we bring our feelings to expression

8m   so do we also learn to apply it to the threefold division of **man – we could equally well take a different scheme as our foundation.** And it becomes possible if we have here **[Plate 1, middle]** Saturn, Jupiter

9t   I shall now be able to place a similar plane perpendicularly through the Sun. **[A vertical is now drawn.]** Thus the movements

10b-11t   **That is connected with certain things that I mentioned here in the public lecture day before yesterday.** The anthroposophical **worldview leads** matter, the material, **into** real spiritual knowledge. ... Man does not **put himself to the test** thoroughly enough. He does not ask himself about the true nature of what these words designate. A great deal has become **a mere cliché.** ... But our speech becomes **intuitive** the moment we pass from the mere making of **clichés,** the using

11m   With the right sense-organs we '**feel our feeling**' of them, as it were. It is then that the first become our own. We could never have attained **a representation of the** ego ... By simply laying our hands one over the other we have a picture of **the representation of the ego.** It is indeed true that by beginning to use **intuitive** images instead of mere **clichés**, man will become inwardly richer

11b By taking this path we shall find that the universe comes to life again for us, and **a feeling** that we human beings **participate** in its life.

12t By simply stating their distances in abstract numbers – **and in the end everything that has to do with contemporary astronomy is given in this form** – we have said nothing

12t this organ may be a certain distance under the line, and another organ a certain distance above it **[Plate 2, in the middle beneath]**; it is not, however

12b think of the universe as an organism depends on our learning to understand the **hieroglyph of organizing** we have before us. We must learn to perceive man as a **hieroglyph** of the universe,

13t we learn to perceive scientifically the connection between moral law and natural necessity, **so that these no longer diverge**. Today I have

13b which we so need in today's society.

We shall speak more of this tomorrow.

Lecture 2

14t **We want to continue with yesterday's consideration**. Yesterday I drew your attention

14m We certainly speak of length, breadth and height, but if we **think** our three planes in this way

14b built into the human body becomes quite abstract when used by man to **construct in intuition** the whole universe with the sun and the stars.

15m the head alone but chiefly functions there, and extends throughout the rest of the body **in all its branches**. Then there is

15b We now need **to take really seriously** this threefold man. **We have to imagine ourselves schematically** in terms of head, rhythm and limbs. **[Plate 3, middle]** Of these three, only the **'limb man'** with its continuation … of our earthly planet. **The 'limb man' is strongly integrated into the forces of our planet earth**.

16b stem, and during the year forms its fruit and seed. **[Plate 3, right]**

16b **We can schematically draw this cyclical development thus: the** plant proceeds from its seed

17m unless we take the seed from its own plane or region and return it once more to the earth, **so to speak at a deeper level. [Plate 3, left, above]**

17m we must bring it down to earth once more. **[Plate 3, left, below]**

17b We can now draw the process in a diagram. **After the plant has come this far, it can go no further.** If this is the earth level here [horizontal], then the cycle of evolution for the plant must be drawn thus [Plate 4, left]. But the plant must again

18t forces are able to act freely. The bindweed is an instance of such a plant growing toward the spiral. [Plate 4, right, but only the spiral with the old and new seed] Thus it completes its cycle.

18m head is formed [Plate 3, left] in the same way as the new seed

19m advances, moves forward, while the head rests as though in a coach, not taking part

19b celestial influences. Man grows downwards. He has that which withdraws itself above, and everything that grows into the earthly influences is what grows downward. When he arrives

20t gradually becomes subject to earthly influence. What has given us this possibility? What is the explanation

21t This will show you that man, like the plant, is embedded in the forces of the cosmos; but with this difference –

23t movement of the earth taking in its purely absolute sense. [Plate 4, right] If, instead of following the motion

23m the earth is here and the sun there. [The positions of the Sun and Earth are drawn into the diagram.] An observer sees the sun in this direction. The earth progresses,

24t of his own movement, the impression that the earth does not run around.

24b of the *universe*.

We want to speak further about that tomorrow.

25t   must not forget that **the** Copernican **paradigm** arose during the

25m   home to me in the **question-yearnings** arising out of the last lecture. …again, to draw the lines that would give us a **most abstract** picture of the world.

26t   It is not my wish to offer any similar type of **worldview –**

26m   pass to a more intensive, inner experience of the three directions of space **for humans themselves, who experience their own figure,**

26b   Let us consider *colours* once more – **really I want to mention the example yet again. [Plate 5, both quadrants, the left yellow, the right blue]** … gave rise to **the** Copernican **paradigm** has also led us to say

27t   That is really just as if you were to treat the word PICTURE in the following way. **[The word *Bild* [picture] is written on the board.]** Suppose **someone calms and broods over the word:** ' "P", something

27m   makes me **distance myself** from it.

27m   of searching behind the external universe for 'vibrations', the atomists should seek **and seek** for their own ego behind the phenomena and then try to find out how their own ego is **bound by** the outer

28t   seems to pass through the zodiac in its yearly round. **In its daily course also we see it in a sense running through the zodiac.** Now if

28m   blackboard. **[The circle is drawn to the left, above.]** We shall then have one plane

30t   part of what appears to us in outer **worldview**? We emphasized

31m   We might be inclined to call this, therefore, something very much **irregular**. Man seems to

34m   Let us draw it in diagram. **[Plate 5, on the right, the left-hand stream]** Within this natural event

34b   To draw this, we must present it as a different stream **[The stream to the right, red].** Yet the difference … in quite a different way **[in the middle, below, the white nucleus]** … deeper for the second teeth. **[The broad ring around the nucleus, red. Seven sections are indicated.]** If we draw them side-by-side

34b   unseen forces of attraction, or spraying out light to them, and the like **[Plate 6, above, circumference red]**; … The sun is the *result*, not the point of departure **[the same Plate, below].** The living

35m   seven different points, say, as it passes round **[below to the right in the plate, the large arc red].** The simple fact

35b   such and such a line **[middle of the Plate above].** That tells one nothing at all, it is an empty **cliché**. We can only know

36b   and we can trace its path. It is all within us **[the same Plate, on the left below].** If we now go out into the universe and look for the sun, it is the sun which corresponds to the heart within us. **(We will want to speak about this next time.)** What passes from the

38m   You see we cannot get on so quickly with constructing of **paradigm**. In order to make a few of the features of this **worldview** clear, we can of course draw a **screw-shaped line**, but this does not yet show the actual state of things.

38b   that can be done very quickly. The wish to construct a **paradigm** with a few lines must be relinquished, and we must learn to regard our present conception of the world as **a delusion. And the mathematicised world is especially a delusion.**

39b   means of two corresponding bridges.

We want to discuss this the next time.

40t   least one aspect of a kind of **sculpturally complete worldview,** which can then lead us

40m   governing the upper and lower realms. **The difference can already come before our souls if we consider it that our upper nature, which is ruled by the sculpture of our heads,** is in its origin

40b   of course, to the outer **substantiality,** but in regard to its

41t   live between death and a new birth. Here, then, another world **towers into this world.** Another world is manifested

41b   our limbs, *how* the will **shoots into** the movement; we only

41b   So here we have before us an entirely different world. **And if we want to bring this other world, this whole state of affairs, before the soul, then we must say: Here there is somehow a world [Plate 7, in the middle below; red portion of which the horizontal red arrow points left at the red arcs], that reveals outwardly what speaks to our senses. That which speaks to our senses we perceive through our eyes, our ears, etc.** To this world we belong

41b   through our limbs **[blue, on the right red; arrow downwards and descending arcs blue],** a realm in which we are unconscious

42t   origin in it. Man felt in olden times – and the East still feels it – that **mediation** between the two **exists.** ... When we eat, **then it is the blue line that actually symbolizes our eating. The** process takes place in the

43t   Now it is a fact that the process of breathing is also a kind of **metabolism,** at all events ... state between actual **metabolism** and the

43t other, different world I have spoken of, **which I have represented in this scheme as blue [the previous drawing]**, and the dreams reveal

45b earth's axis **[Plate 8, left, above]**. The earth dances, and in

46m one year and one human respiration; we have the second world **towering into** our own.

46b another world **towers into** this one, and is regulated by the

47t different laws **[Plate 7, on the right, in the middle, diagonal shading blue, horizontal yellow]**. … thread of cause and effect, so long shall we be shrouded in **dreadful** illusion.

47t earth's axis that another world **towers** into this one – only then

47t the soul-world and the material. He who can **observe factually** what is contained within his own self

48b is something less than empty space **[Plate 7, bottom right, still without rays]** – a sucking, absorbing … with any product of combustion in the sun, but is a reflection **[the rays are added]**, a raying back

51t day astronomy calls 'precession', **the advance of the equinoxes**, we have something … a third world. **[Plate 7, drawing on the right; in the middle, second horizontal shading red]** We have thus

51b We generally speak of moonlight as being reflected sunlight **[Plate 8, beneath red; the small circle on the left blue]**. I am speaking

52b Jupiter, Mars, Earth,, Venus, Mercury and Sun **[Plate 7, below on the left red]**. That is not what is wanted at all.

53t connected with the rest of the **sidereal system. The** moon, however, is related to quite another world; **it points to a world that is shoved into** ours, and which represents

53m in nutation are motions proceeding from **astrality**, and not … It is just this **Newtonianism** that has cast us down so far

53b   attraction exists between these bodies, like some invisible cable **[Plate 7, right, above].** But if really nothing but this force

53b   see! Here we have a planet **[the same plate, middle, above]** imbued with the constant desire to

54b   **We are standing within such things today**. It is from ideas of this kind that the solar system is constructed today.

55m   system in its relation to the whole visible **sidereal** universe.

56t   century. This is what Goethe always fought against in his scientific investigations, and against which all true **Goetheanism** must also fight. Goethe **already** found himself compelled

56m   The following example will show you **how little value some** of the assertions made by modern astronomy **have**. Modern

56m   sun attract the earth, but the earth must also attract the sun. **I draw now in a way that is similar to the way in which astronomy draws [Plate 9, above].** This, however, obliges us

57b   world-picture. We **concluded yesterday by saying** that we must imagine, first of all

58b   the regions that comprise the zodiac **[Plate 10, left, above].** I do not intend here to deal

59b   he moves around within the **celestial vault**. A similar gradation

61m   points to the fact that the effects upon us of that **sucking etheric** ball the sun very greatly

61b   will influence us directly **[in the drawing this part is shaded]**, the other not directly, but rather

62t   their non-presence. The fact that something which is *active* within us and **in a sense** enables us **to have the possibility of bringing** into a kind of relationship what is working directly upon us and what is absent, from whose direct influence we are **removed. However, this** opens up another possibility.

62t   influence **[arrows]**; the influence of Cancer thereby gains

62m I want you to note this point specially, and then you will realize that you cannot simply say: Above us we have the signs Aries, **Taurus, Gemini, Cancer,** etc., and below **this and that,** and so on

63t these varying influences, while yet on the physical plane? No, **we said that just a moment ago**; we have seen that

63m through this non-activity, more or less in a state of sleep. **You need only remember that humans actually sleep their way into the world.** Children are always more are less

65m matter, with the upward striving forces, with what grows beyond the earth. **[Plate 10, right]** And we must observe

66t whole quality and nature of man. We find, **when we consider human nature**, in the first place his

67t sense the heart has been **welded** together. By pursuing embryology, we find how the heart is gradually **welded or pushed together**, as it were, by blood circulation, and

67t the idea, let us say we have a stream of water falling over a rock **[Plate 9, right].** It throws up a variety of water formations

69b influences from all other super-earthly influences.

More on this tomorrow.

70t had to be regarded, to begin with, from four points of view. Firstly, in relation to the **formative forces at work within him; thus what gives us our actual human form**; secondly in relation to all the forces **corresponding to the motions of the inner humors** expressing themselves in … have the organic forces; and fourthly, **actual metabolism.**

70m

1. **Structuring forces**
2. **Forces of inner movement**
3. **Organic forces**
4. **Assimilative, or metabolic forces**

71t have a connection with the activity of the senses, **to the extent that the senses perceive**. We shall

71t We must imagine this light-activity is something intrinsically different **when we think how the eye is built up**. We actually arrive

72t the **sidereal** universe then acted upon him in a different way. This **sidereal** universe was also differently formed from the way it is now. **They affected the human being other than the way that they do now**.

74m Yet there is a way in which one can **rigorously** *prove* the motion, … are no other proofs but inner processes. **Accordingly, we have to be able to point to inner processes if we are to speak of the absoluteness of a movement.** Applying this to the earth, we can truly speak there too of absolute motion, for **anthroposophy** enables us **gradually** to realize that this motion

75t   When we look at the zodiac, it is the outer, cosmic representative of our own outer form **[see schema]**. When we consider … between (2) and (3) **[see schema]**, between the inner

75m   in the course of their **vascular** flow

75b   whole body. But this is not the case; on the contrary, **the blood is something inwardly mobile in itself and it has its vitality**. The pulsation of the heart … are integrated into **the flow of living motions**.

76t   motion. Were we to study the phenomena of nature **that take place generally in our environment** more deeply … We shall realize the harmony existing between **the phenomena of the weather** and the motions of the planets

76b   the same way as metabolism is related to the course of the day. **Metabolism is related to the course of the day**. Observe how

77b   compels us to speak of a rotation of the earth on her axis **[Plate 11, left]** causing an apparent motion

78t   year **[Plate 11, on the right]** but rather with a movement of the Earth in the course of the year which corresponds to the solar motion. … between the two. **[Plate 12, left]** This is something in Copernican theory that will have to be substantially corrected **in the future**. But there is

78m   Only part of this process occurs **in such a way that it is accompanied by the appearance of consciousness. Another part is accomplished** while

79t   like – is the lemniscate **[Plate 12, middle]**. Here you have … in the other direction; **once in the direction of the one arrow, another time in the direction of the other**. The directions alone are opposite, the two points being equal conditions **with reference to the vital relationships**. We can now

79m   motion of the earth, we can no longer attribute a nearly circular motion **to a single point. We may not do this. Rather, we have to** imagine that the motion is another. **When we fasten upon the actual motion of this point, we obtain such a line. [The lemniscate is drawn on the globe, Plate 12, right.]** The motion is not a simple … which is also the **representative** line described by the metabolic process.

80t   relative or actual motion! **In a real movement we cannot say**: Perhaps the movement

80m   a yearly motion of the earth around the sun. For the earth follows the sun **and both execute the same rotations**.

Various other facts

81t   found. It is a **kind of** screw-like path … human beings. **What kind of a motion is it?** What can we find

81m   the forces of form; but one which must be of long duration. **The Sun moves somehow in relationship to the remaining fixed stars.** The forces of inner movement

83b   the rhythm of our daily life. The **formative forces** have solidified. … 'Animal Circle', not of the **'Human Circle'**, as corresponding

84t   the zodiac is **the substantiality that** corresponds to our ego.

84b   takes place by way of movement in time in this **sidereal and planetary** heaven, … *outside* this time, outside **even of** all that takes place … past. But the past does not prejudge **in the cosmos** the whole of the future, it

85t   *in his moral actions.*

Let us hold fast to this point. I can only lead you forward

86t   Tomorrow I shall go over again some of what has been said in order to then to connect it with something fresh. **Today I will add to the course of our considerations something that is suited to indicate something more specific about getting to know the cosmos.**

87b   greatest, judging by modern methods of thought? **Just consider.** It states in a grand manner

87m   then creates man as a composite of all that is found there; in other words, it examines everything **extra-human, in order to say in the sense: Here, at the endpoint the extra-human ends. Now it is a matter of human beings. Humans themselves are not considered.** Natural science does not feel called upon to study man as man, and consequently any real understanding of his nature is out of the question.

You see, here I would like to proceed from something that I developed at an entirely different place, in an entirely different context, before a different public, but it can have an enlightening effect in this context.

87m-b   In fact, people who claim to be experts in this domain of nature nowadays **really have a very great need** to examine Goethe's investigations in natural science, particularly his theory of colors. He uses **an entirely** different method of investigation from the one we are used to today. At the very outset he mentions **so-called** subjective **colors**, physiological colors; and what **the human eye as something living** experiences in connection with its environment **is** then carefully investigated.

87b   connected with this. You gaze at a red surface **[Plate 13, rhombus on the left]**, and then quickly turning

88m after-effect comes forth in the form of a **memory** picture. The whole phenomenon

88b memory picture that rises up within us in relation to a certain experience? **We want to leave totally indeterminate whence**. As soon as we put our question in this form and require **an appropriate** answer, we realize that the whole method of modern scientific thought completely **fails; and it** fails because of its ignorance of one great fact – the fact of the universal significance of *metamorphosis*. **The whole of the universal significance of metamorphosis is something that contemporary science does not know**. This metamorphosis

89t You will remember that in order to gain a true insight into the whole nature of **man – we may,** for the present purpose, consider the last two **systems** as one – and we then have

89m a complete transformation **in the cosmos** in the period between death and a new birth

90t of the changing impressions are transmitted inwards to the power of **understanding.** In any particular organ of the head … We might depict this fact diagrammatically **[Plate 13, above].**

90b that the organs of the body (taking 'body' as the opposite to 'head') **metamorphose themselves by turning themselves inside-out**. So that one or other of our

91t become an eye, and has attained the capacity to generate after-effects from the external impressions it receives. **This capacity to call forth after-images of the external world must come from somewhere**. Now where does this capacity originate?

91m Now compare this with the inner organ of which the eye is a metamorphosis in respect to its activity and forces. Let us look at the organ in our inner organism, out of which the eye developed. … means to remember it. So when we look for the **original** metamorphosis of the eye's activity

91b *within* the **body cavity** which correspond to what the eye

92t through the eye. We *hear* sound with the ear, and **with the organ in the body cavity** corresponding to the ear we *remember* that sound. Thus the whole of us becomes, as we direct or open our organs inward, an *organ of memory*. **The whole human being is an organ of memory**. We encounter the outer

92m comprehend matter, because it sticks fast to its abstractions! **I already pointed this out**. It becomes more and more abstract, **that is to say, more filtered, spiritual;** therefore it cannot

92b whole body. In this connection our science perpetuates **an asceticism, a continued, one-sided asceticism. Wherein does this asceticism consist? In not grasping the material world in its spirituality; rather, they want to despise it, to overcome it, to have nothing to do with it.** Our science has

92b the entire remaining organism! Here it is that memories originate through the **recursive oscillation** of the organs.

This was very well known when no spurious asceticism burdened human **views**. Thus we find that

93t that hypochondria is merely **psychic**, is something

94m of injuries produced upon our **organs. We** receive help

94b man and microcosm was **sought** which we find

95t transmutation of matter from the outer world. **We inwardly participate in this.** This day-cycle within us

95b times that of the other, returning once again to Sunday [**Plate 14, left**]. We experience this … motion of the sun. **We call it apparent today, but at the moment that does not concern us.**

96m impels us to reckon, now with the day-cycle, and now with the week-cycle? **Whence this feeling?** It arises from

97t organism. Let us translate this into rational language. **We have explored this in a real language; now let's translate it into a rational language. You have eaten today.** Let us say that today

98m Goethe had **a magnificent** intuitive sense

98b Imagine we have here **schematically [Plate 14, right]** the upper or

102m   attention to the **comprehensive** metamorphosis that takes place

102b   birth and death. But the content is essentially different in each case. **If we want to make schematically clear wherein the difference consists, then we can say: If the human being is here [Plate 15, middle, above] between** birth and death

103t   *the center* **[Plate 15, right, above].** And then we know

103m   It is the fullness or poverty of the content that matters, not the size. **Just that must come to our consciousness again and again, that it depends upon the wealth or impoverishment of the content.** If we combine all we observe in the mineral, plant and animal kingdoms, **in the realm of the forests and mountains,** and add thereto … But then the spiritual world takes up the formative forces of the **remaining organism**, which as inner **experiences** belong now to the periphery

104m   molecules which are composed of atoms **[Plate 16, left, above]**, that the atoms … by the *whole starry heavens*. Suppose here is a planet, here another, here another, and so on **[little circle in the drawing, on the right]**. Then there are the fixed

104b   *reflected* in the molecule. Instead of **peering** into the infinity

105b   object, it attracts it and draws it into itself **[Plate 15, left, above].** The activity of

106t   the absorbing action of the **etheric matter or** ether. Your nose … appearance of your front and back. These are things which are not usually taken into consideration; **in the materialistic age we do not think of them at all.**

Further, when we come

106m   the things that must be taken into consideration **[Plate 15, middle, below]**.

If the ether has merely

106b   we find in another plane if we have a bottle of soda-water **[Plate 16, left]**. We may not see

108m   objects here. Then we *live* in this terrestrial motion; we begin to **participate the motion of the Earth.**

There are also other motions connected with the starry heavens, all of which we participate in **[Plate 15, right, above]**. Correctly considered

111t   ordinary conception of the **solar and** planetary system ... celestial bodies **[Plate 15, right, below]** – so than any interference

111m   be fastened **[drawing]** (where or how

113b   From **the most various considerations in which I have engaged** you all know of the relationship in human life between **the daily rhythm of** waking and sleeping.

114t   state of sleep. We must be absolutely clear that**, even when** we are awake, this state of sleep continues as far as our **will** is concerned.

114b   Thus **it is actually a given fact that** man, in his life between birth and death, is in **a fluctuating** waking state … We must realize at this point that really to understand human nature **presupposes that we fix** our attention upon the fact

115t   the will-nature or metabolic-nature, we do not **understand** only **schematically** the outer limbs,

115m   erroneous hypothesis! For spiritual observation shows the following: **If we take the human being schematically [Plate 17],** when someone's will allows him to raise his right leg, his ego-being, **his real ego-being,** exerts a direct influence … nothing of it. **The nerves that are inserted here, that then proceed to the central organ, merely inform** us that we

117t   limbs, is associated with Mars. **[The Moon and Mars are added to the drawing in Plate 17.]** People of today

117b   found introduced into quite modern literature **by mystical fops,** then they **need all the more to be written off.** Up to the earlier part

119t   only the other side is turned toward us, **all according to whether you have illuminated it from your upper human being.** It is exactly the same with the moon **[Plate 17, right].** She revolves around

119m   more we find this to hold good. It was the **naïve,** instinctive observations … these inner relations between human nature and the celestial bodies, **while today's materialistic philistine says: Well, yes, silvery light. For the similarity with the silver in light both have taken the same sign. All of that is nothing other than ignorance of that magnificent knowledge that was not attained by the ancients in the way that we must gain it again. They attained it in a different way.**

Now let us take

119b   limbs. Hence we find that a person **moved by elementary** feelings generally accompanies

120t   circles – I will not name them – it is considered a sign **of a serene nature** to abstain … putting his hands in his trouser pockets while speaking, it may not only mark him as a man of **serenity**, but also

121t   instruction is given in handicrafts **by our friend, Ms Molt, together with several other ladies**. You would see the boys as well as the girls **diligently and devotedly** absorbed in knitting or **crocheting.**

121m   to cross the middle finger over the first, **like a caduceus**, or not. **This makes a great difference to our thinking**. The movements

121m   acquire facility in crossing the middle finger over the first **elastically,** making a serpent entwining the **caduceus,** but it is

122t   abstract, but in specific **psychic and spiritual** terms, which can have a real effect on all areas of our human existence.

One cannot advance in these things without **also** constantly **having recourse** to the other side

127m   receives **from** the super-earthly worlds

127b   extending beyond our earth-related organization, of **towering** beyond it with our head

128t   the result of the foregoing incarnation **[Plate 18, left]**; while the will-forces

128b   These forces must first be stemmed **[the lower arc is drawn]**; they must make a halt

129t   until sufficiently filtered, **sufficiently diluted**, given more … will-forces are arrested in the organization of the *larynx*, **when they shoot up in man**. In the male organization

129b   way soul and spirit manifest, **reveal themselves in the outer, physical world**; and at the same time

130m   So you see how one can gradually come to a real investigation of the mutual relationship between **psychic and spiritual** and physical, **bodily** nature, and how

131t   expression of something that takes place in him organically. If we **properly** study the metabolism of the animal **anthroposophically**, we find that … to a certain stage must in man be arrested at an earlier stage. **If I may draw it schematically [Plate 19, to the left above, the strong horizontal stroke with the spherical end], that which is driven in the animal to a certain level, must stop at an earlier stage in humans.** Superficially expressed

131m   revealed when **anthroposophy** is allowed to eliminate

133m   today. For what was the **worldview** of our forefathers – that is, of ourselves in our former **incarnations**? What was it?

The **worldview** that man had in those times

133b   human being. **That is what gradually awakened in humanity as it developed further.** In primeval times

134t   cosmic picture **[Plate 18, right]**, and it had

134b   Ancient **Hebrew** Mystery teaching spoke much

136b   times looked into this outer world. **Let's say this is the boundary of his senses vis-à-vis the outer world [Plate 19, to the right, above].** By means of all

137t   he carried it in his senses. **[The two little circles on the right are drawn.]** This sense for reality

137t   training we receive in **anthroposophy**; only then shall

137m   arising no other way than by developing it within us **through anthroposophy** first.

137b   being through **anthroposophy** in a manner in which the outer world … Hence the need for **anthroposophy**; for people cannot … do so by means of a training in **anthroposophy**. In former times

137b   world. This is now no longer understood, nor can it ever **be understood in its origins** if the foundation,

138t   clairvoyance the facts of **anthroposophy** are investigated; … *conclusions of **anthroposophy***, not to satisfy a curiosity

138b   would be sundered **[Plate 19, very top]**; and this they cannot be.

139t   line, a lemniscate **[the same Plate, middle, below].** When we say

139b   and so attain a *progressive* lemniscate (A) **[right, below]**. Thus when we observe

140t and continue the development, we obtain **a screw-shaped line**. This **screw-shaped line** is **intimately** connected with our

140m of Mars as moving along any other kind of line then this one **[left, middle]**. Gradually … continue along this kind of line. **[Steiner points to the line just drawn.]**

140b Instead they took this line and set circles upon it **[left below, the great arc]** (B) and

142t   cosmos outside him, **to the extent that a definite relationship to the extra-human cosmos exists**. And now I would like

143m   that is to say, representing it in thought **[Plate 20, right, above]** as a line in which

143b   such a way as to **say [the interrupted straight line at intermediate height, from right to left]**: Today I

143b   daily cycle comes out as a **lemniscate**, with the point of awaking

144m   and the etheric body; *that* lies in bed. Strictly speaking,

144b   our **personal** winter, and that between falling asleep and awaking is our **personal** summer.

145b   with its **apparent** 30-year orbit of the sun **[Plate 20, left, above]**: How can we express

146m   mucilaginous organism **[Plate 21, left]**, itself revolving

148t   in a similar curve **[lemniscate, Plate 20]** – only, of course

148b   observed merely as a body traveling about in cosmic space; and the same with the other planets. **(I must again present something in somewhat of a summary fashion, which I will explain more fully later.)** … space. Saturn **pulls** our system in space, embodying that

149t   lemniscate **[the previous curve]**, but when we look at it sideways, we obtain lines which are continually rising upwards; there is a progression **[Plate 20, right, below]**. This progression corresponds

149b   first of all necessary to gain **for** inner **reasons an** understanding of the Earth's **lemniscatory** movement. … And now I come to what I wish to point **out. Astronomy tries to draw this movement.**

Astronomy wants … system and explain it by calculation. **But planets** such as

150t   and Mercury, however, **also** have relation

150t   and Mars, we come at **best** to a boundary,

150b   solar system in terms of three-dimensional space alone **[Plate 21, middle]**.

151t   **All these planispheres** and so forth we have to look at as follows. If here we have Saturn on the **planisphere** and **somewhere according to our usual schematic solar system** Mercury, then it is

151b   vision of **anthroposophy**! It is of urgent necessity

152t   planisphere showing Saturn and Venus in the same space. For **this anthroposophical** view of the universe which we are giving here is not merely mean that we hold something up before us, but also that, in a sense, we *learn to think*. What exactly does this mean **when we learn to think like we think today**?

Remember what I have said: when our bodily organization is remodeled **for** the next incarnation, it not only goes through a **transformation**, but is turned inside out.

153t   oppose it by taking up **anthroposophy**. One might even define **anthroposophy** as something that induces us to take our heads seriously once more! From one point of view the essential aspect of **anthroposophy** is really … **Anthroposophy** must disturb the sleep. …

One cannot talk of **anthroposophy** in the way

154b   is preferred by the religious denominations to **anthroposophy**. That is, both Protestants

155b   We must take this **in the fullest sense** seriously, otherwise

155b   fact which apparently unfolds entirely in the **streaming of the** spiritual world, but which

156m   given in *Occult Science*. **We can say: The Ancient Indian was constituted thus, and so forth**. Thus we can describe

156m   occurring during this period. **We can even see that within history**. Let us recall

156b   people of the time – the Greek, Roman and **the Latins**. Let us

157t   The event of Golgotha came **like** an intervention from other worlds. This fact is not sufficiently considered in modern times several historians have alluded to this – **I have already made mention of it**, but they have not

157m   remarkable men – oddly enough, clergy among them – have attempted to explain **the event of Golgotha causally**. **Such a strange person is** Pastor Kalthoff, for instance, **but there are** many others.

158m   abnormal consciousness. This is no isolated case, **but rather others have also sought to explain from the standpoint of contemporary psychology, simply to explain the peculiar form of madness that has come into the world**. And these are phenomena

158b   never occurred. **Anthroposophy** must endeavor to bring

159t   most telling significance **[Plate 22, red; likewise the spans of time to the right and left]**, speaking to us of all

160t   side of the moon in fact comes over the edge, as it were **[Plate 23, above, the sickle red]**, so that really it is not quite

162b   pagan evolution and Christianity; and now, **'ethericity' and 'astralicity'**. Always a differentiation

163t   usual today in people's efforts to **style** everything **in accordance with the materialistic biogenetic law**. The consequence is that

163t   clocks **[Plate 23, below],** one always a little slower

164t   were, generates independent systems or structures of movement, **and beside it the stream of the Sun, that also has independent structures of movement**, so we must realize

164b   evolution – our natural science is still heathen – and **a becoming of man that is of a Christian nature**. In our day many people

171t years. **That is the path which the stars traverse and the sun traverses [Plate 24, above]. But the sun remains a day behind in 72 years**. Multiply 72 by 360

171b Egyptians originally took 360 days for the year? **There a remarkable misunderstanding arose. There was a remarkable misunderstanding that the priests spoke of as a great world-year**. In the cosmic year

175b man evolved **[Plate 25]**. But this is not so. If we take man is threefold, his head alone has evolved from a lower animal form **[Plate 25, right]. What the human being has as head is built up from lower animal forms**. What is

176t This I say for those who **have already occupied themselves with** the theory of descent.

176m in human evolution, and so that you may see the necessity for **anthroposophy to illuminate** all the different realms of

177m   reality has so far passed into the human **souls** only to a very limited extent. The point is, as I have often **emphasized, one that we must fasten on precisely with the eye of the soul.** We have now come to a time

178m   one who directs our attention **very precisely** to the scientific worldview

179b   calling to him from a distance: 'Nothing comes of nothing!' **That is indeed the motto that recurs at the head of the original treatise of 1842 by Julius Robert Mayer: 'Nothing comes of nothing.'** Visiting his friend one day

181t   axles! **Rather, the secondary effect is the primary matter!** Let us reflect, however

183b   while the human spine is vertical **[Plate 26, left]**; and although

183b   to the starry heavens, **which we want to draw schematically, let us say as the zodiac with its astrological signs [Plate 26, right],** which move

185b   lasts the time the sun takes to retrograde 1 degree. **(72, if certain elemental occurrences do not intervene, through which man becomes less old or older.)** Why is this?

187t   science acknowledges took place in us – **if I draw it schematically [Plate 27, entirely to the left, without the hatchings on the head]** all sorts of things

187b   material process breaks down and destroys itself **[the hatchings are drawn into the head]**, and that in

188b   certain speed **[Plate 27, on the right, arrow pointing downwards],** i.e. a flow

188b   going not in the same but in the opposite **direction [arrow upwards]**. The two streams

189t   into one another **[middle of the drawing, on the right]**. If we

189t   One stream whirls downwards **[middle of the Plate]**, and because the other

191m   approaches humanity as a cardinal one, is not peculiar to **anthroposophy**. In order to show exactly

191b   are given by **anthroposophy** today. But all that

192m   modern **anthroposophy** – and demanded by it

193t   ascending evolution **[Plate 28, middle; line bearing arrows]** among the superficial

193b   down into the depths. Only in this way do we progress **[arrows upwards, red]**.

194m   nineteenth century did. These materialists of the middle of the 19th century – **I presented facets of their work in my public lecture in Basel** – actually carried the materialistic outlook to its logical conclusion by saying: If **naturalism** is correct, then there

195t   which had **something rationalistic** not merely for humanity, but **something rationalistic** for all life on earth.

195m   century before Christ. **Until the eighth century BCE**, man did

195b   belonged to the whole starry heavens **[Plate 29, left; arching sky blue, rays yellow]**. Strange as it seems

198t   when, though attacked by Catholic and Evangelical confessions, **anthroposophy** asserts that the Christ-concept and the cosmos-concept must be united. '**Anthroposophy**,' says these confessions

198b   held up as truth! Earth, infinite space, stars **[Plate 29, extreme left; circumference in part on the other Plate]**. Among these

199b   the solar and planetary world, and the earth-world **[these realms are drawn into the drawing on the left, Plate 29]**. We have placed

200t   aorta into the organism **[on the right of Plate 29]**. We can

202b   fact that I **have a certain motive** to give three lectures on the philosophy of Thomas Aquinas at Whitsun. **On Saturday, we will speak about Augustinianism, and on Sunday about Thomism as such, about the being of Thomism, and on Whitsun about Thomism and the present.** I do not know whether

202b   from Dornach! We will wait and **see. But** perhaps it is well

205t One **would** indeed arrive at many things by testing both the logic and more especially the reality of what we find in modern, **so-called** science.

206b space **[Plate 30, large form in the middle above]** as though it were

207t arm as a sort of solid lever appliance **[the same Plate, corner above on the right]**; and of course

207b Taking first the whole rest of man, and then his head **[the same Plate, entirely on the right]**, we can, of course

208t human head is to make all substance fall away like a sediment and be expelled so that nothing remains of it but mere *picture*. ... you have a vessel containing a solution **[extreme left]**. ... above remains finer liquid. This is also the case here **[extreme right in the drawing]** with the human head

208b work in a truly **anthroposophical way**, we can

209t man **[Plate 31, above]**. *Before* the Mystery of Golgotha

209m zodiac. The twelvefoldness of the universe comes to expression in the life of man **[Plate 31, right]**; and we may say

211 or core. Thus in contrast to that first picture **[Plate 31, left diagram]**, which was essentially the Grail picture, must be said the Parsifal picture **[Plate 31, right diagram]**, in which

211m crystallize out in man after all materiality, **substantiality, inwardness, and being** had filtered away

212t   as follows. **[Plate 30, above, to the right of the middle]**. Here is the earth

212t   matter is gradually destroyed by man **[below, to the right of the middle, Earth with lines radiating downward]**. When, some day

212m   earth will be pictures **[triangular forms]**. … evolution through the Mystery of Golgotha **[circle MG with the line radiating into the images]**. Looking towards the end

215b   *upwards* into abstraction, we have thoughts up there **[Plate 30; Steiner draws on the left above]**, and we look

216t   for human instinct, a physical as well as a soul aspect. **Instinct however is at least not the organ of the warmth that we can develop** for our fellows morally

Plates

Plate 1

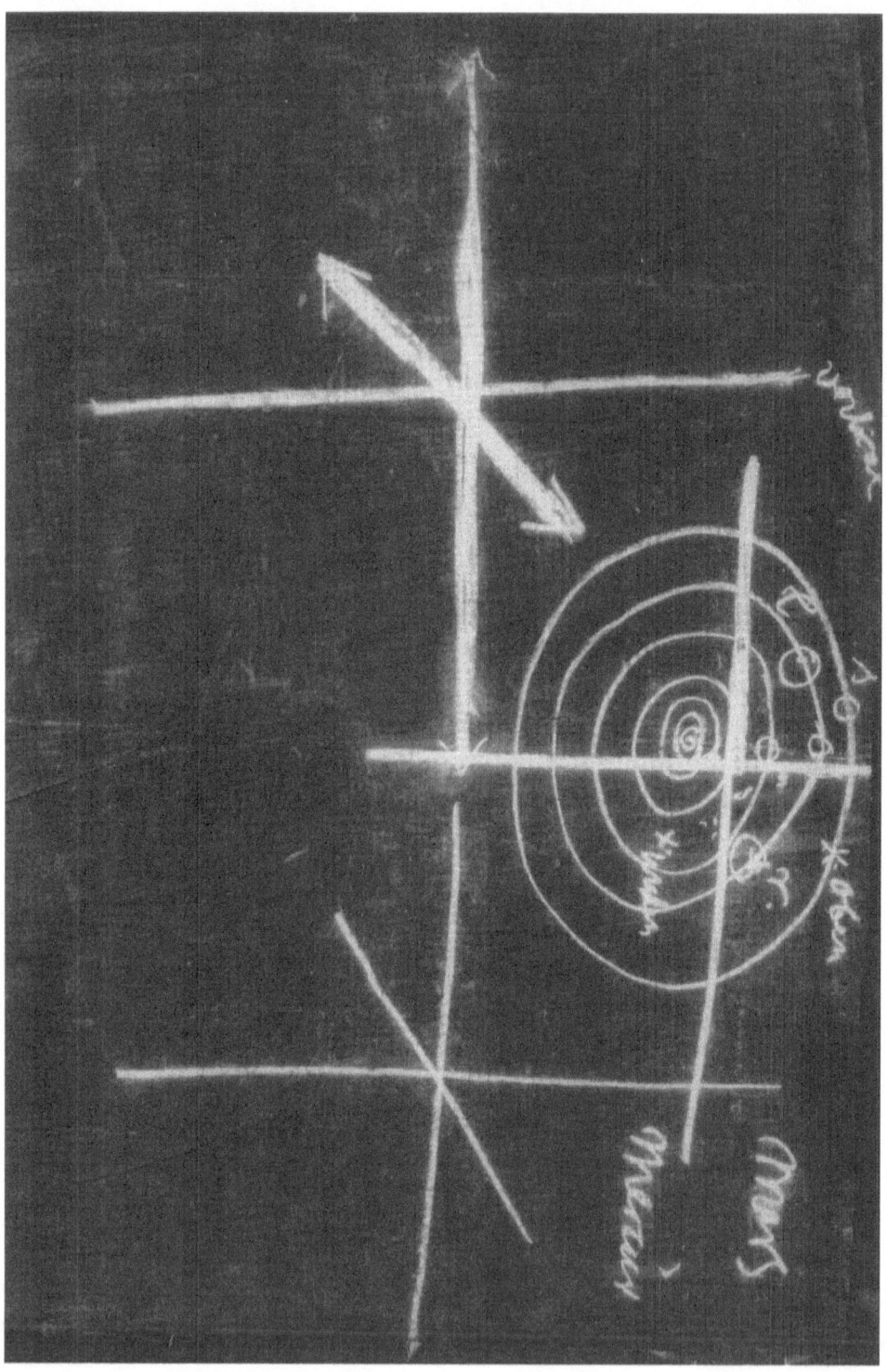

Plate 2

Plate 3

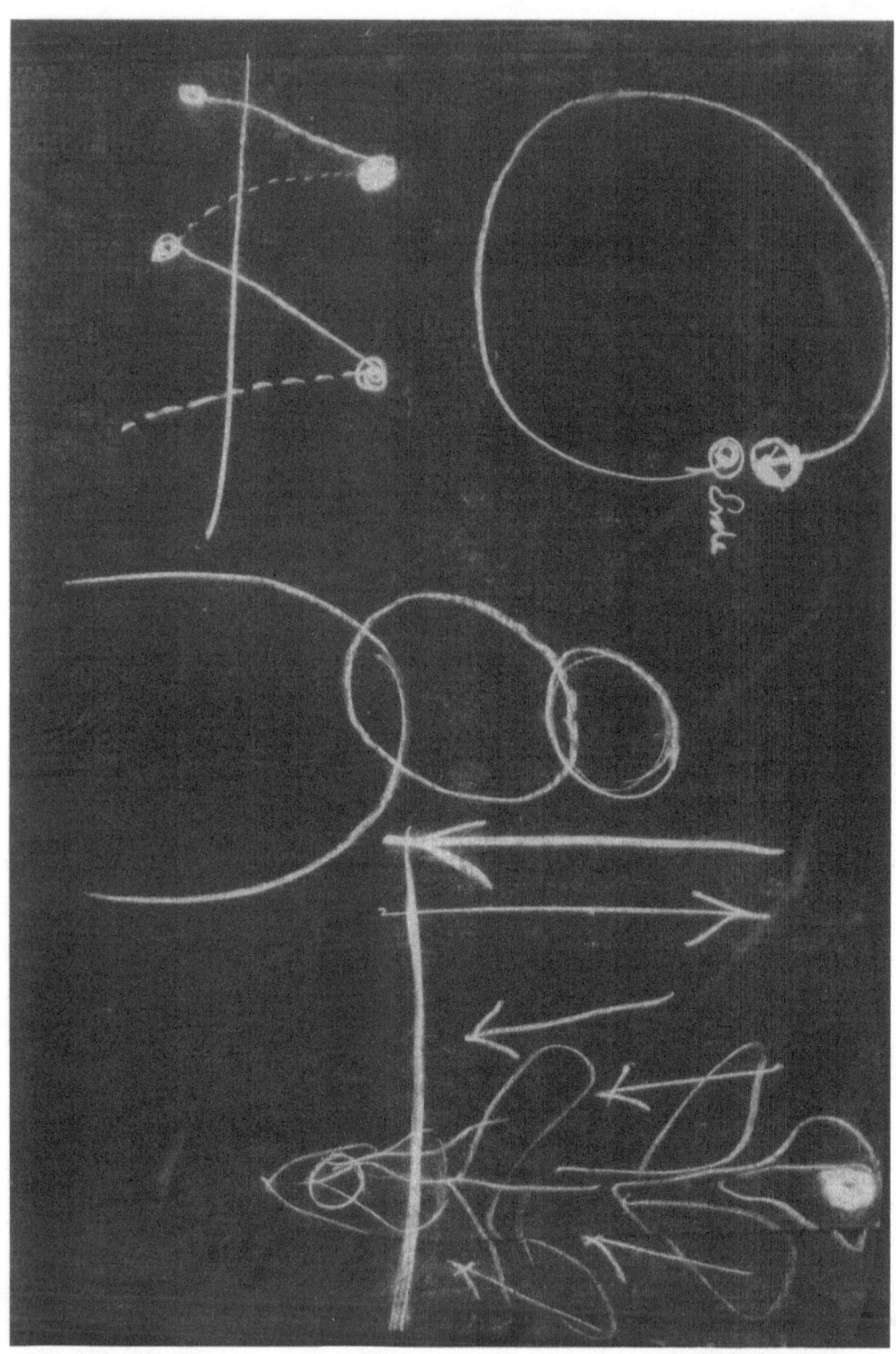

Plate 4

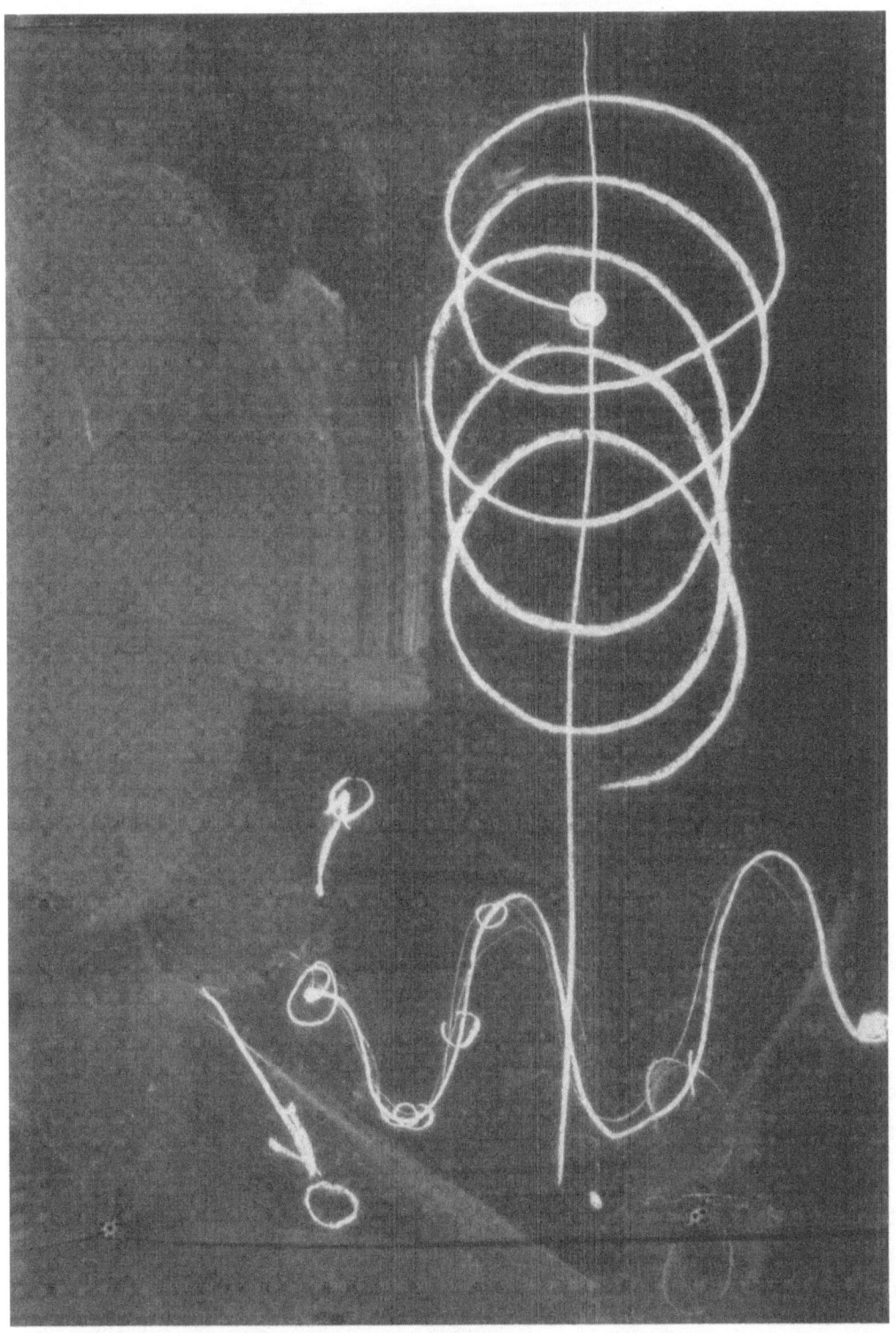

Plate 5

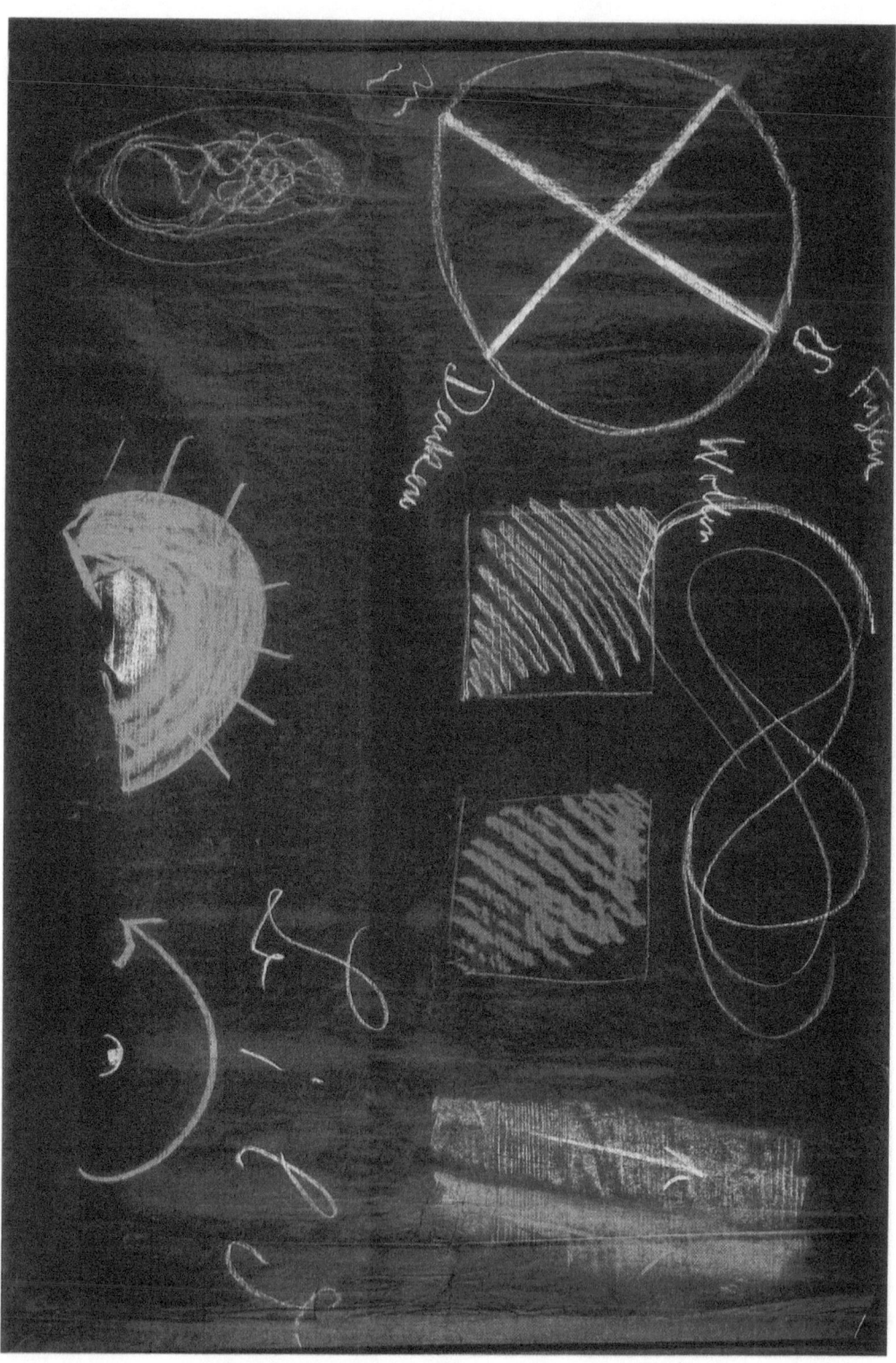

Plate 6

Plate 7

Plate 8

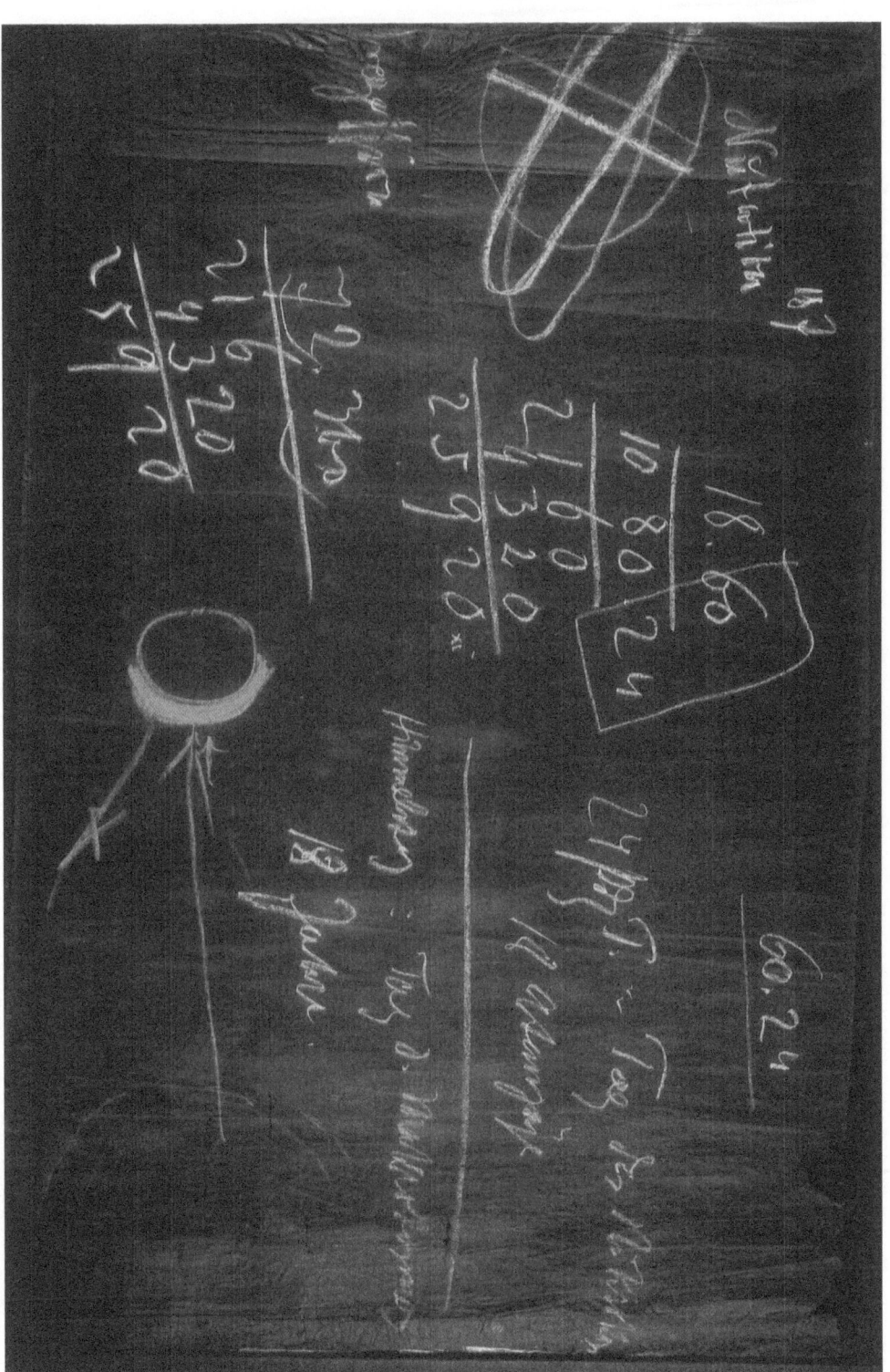

Plate 9

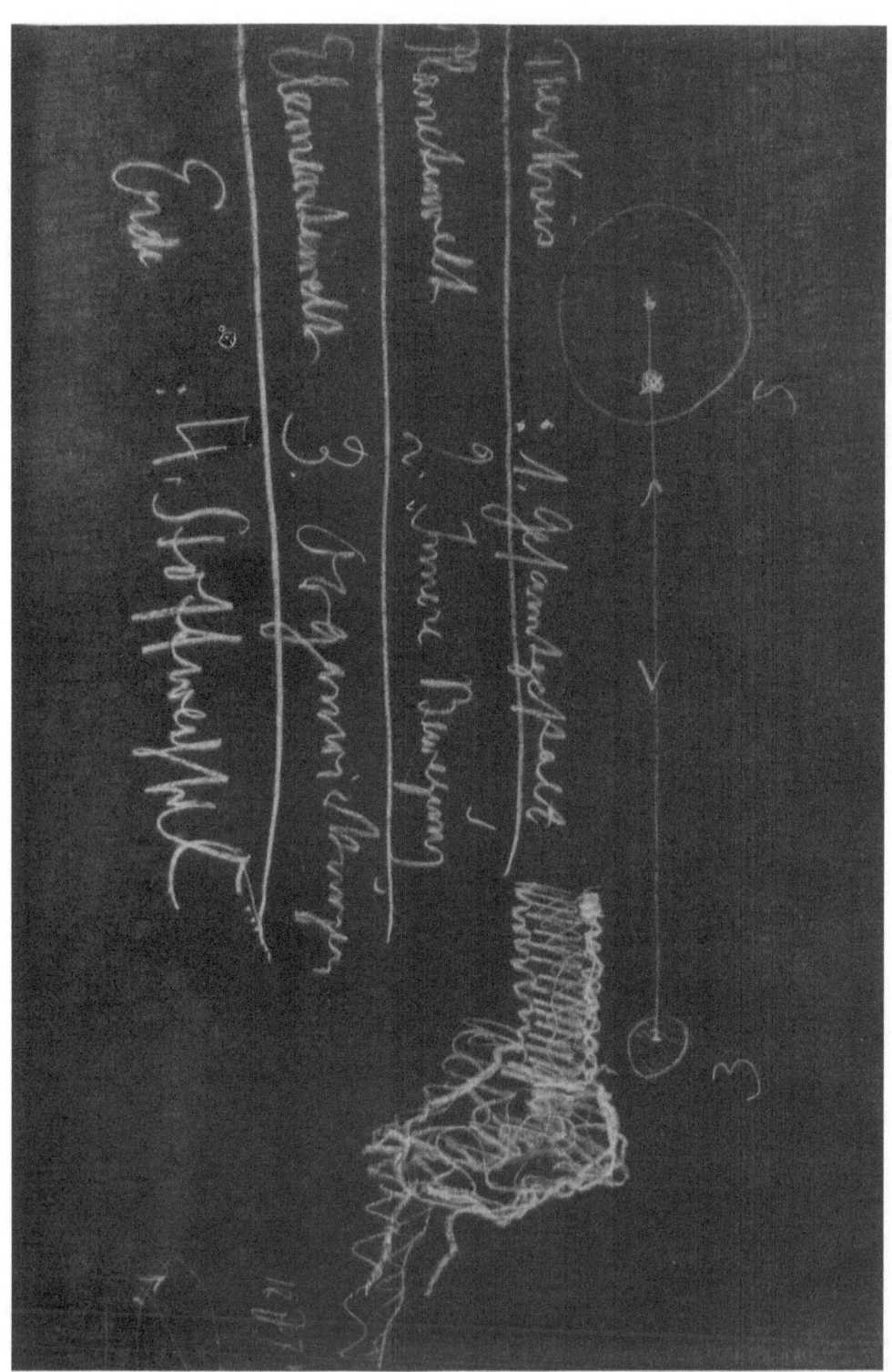

Plate 10

Plate 11

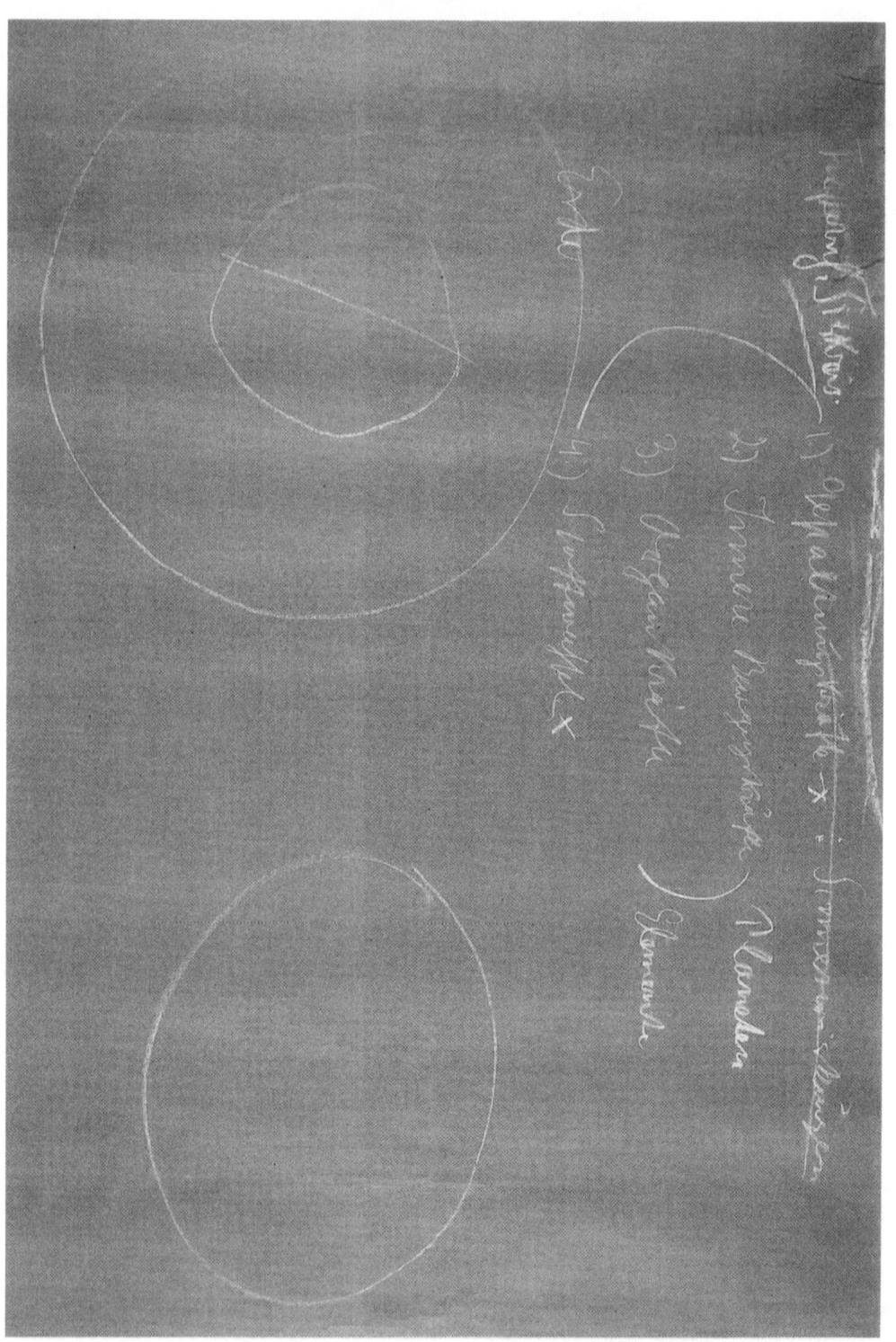

60

Plate 12

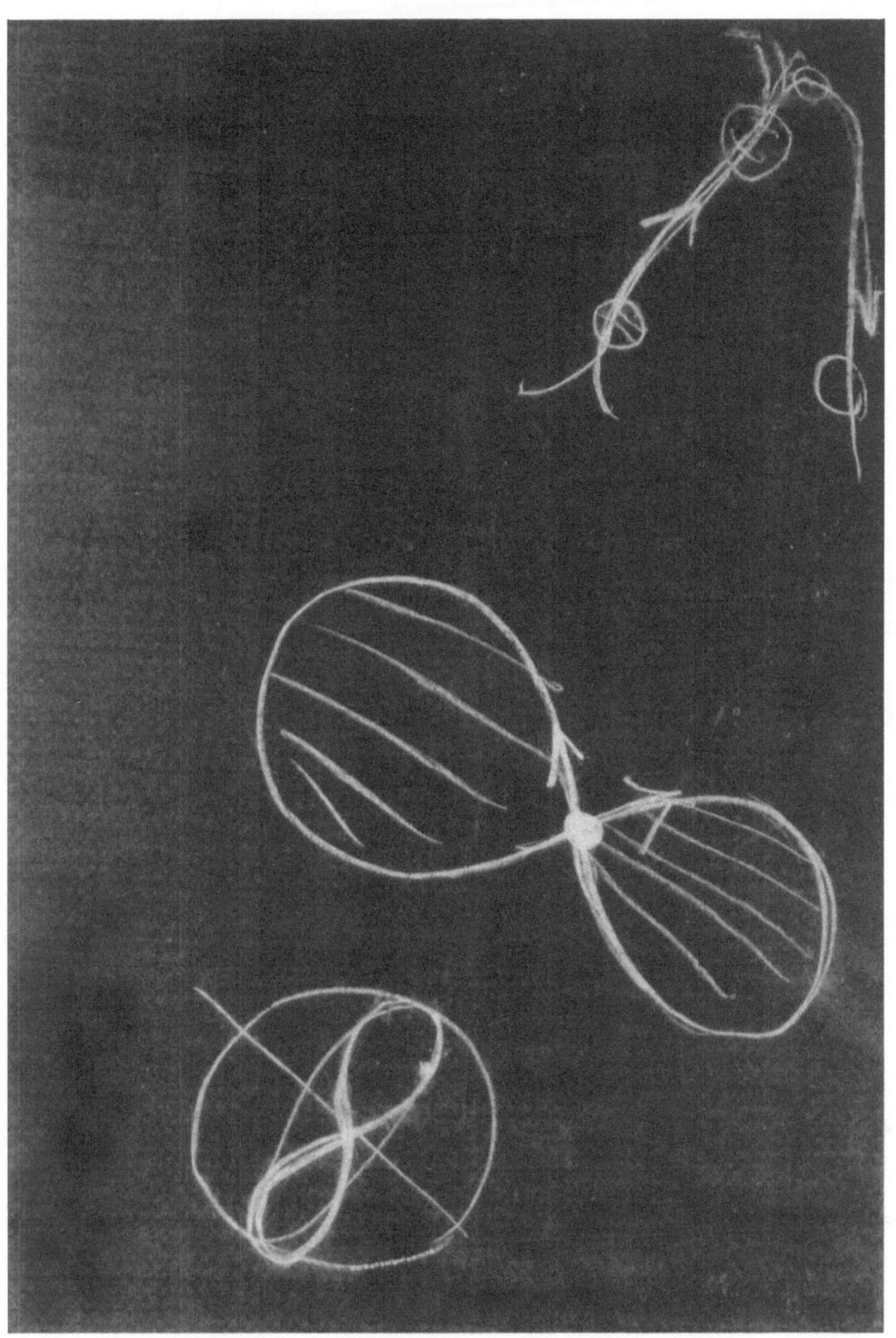

Plate 13

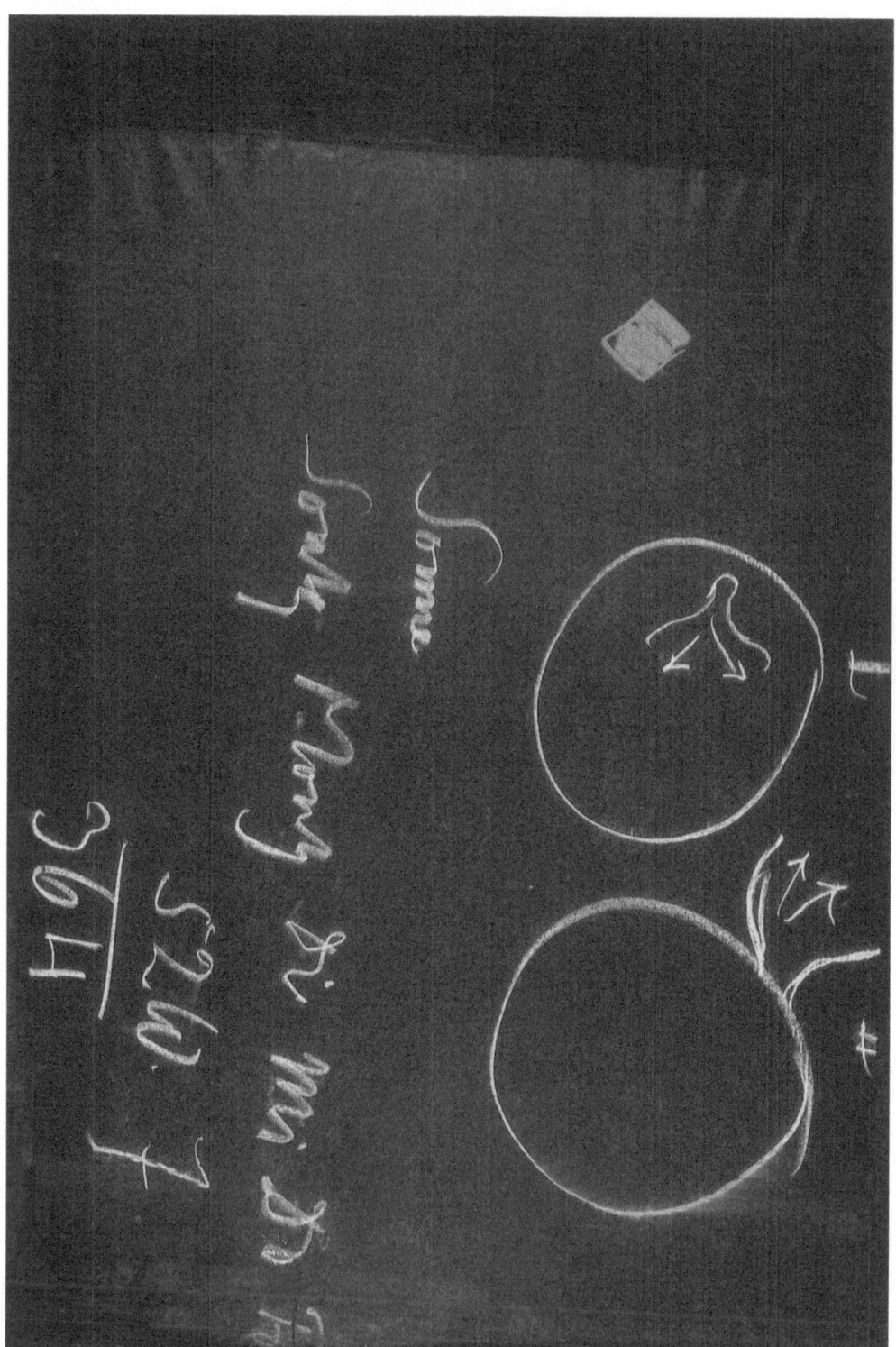

Plate 14

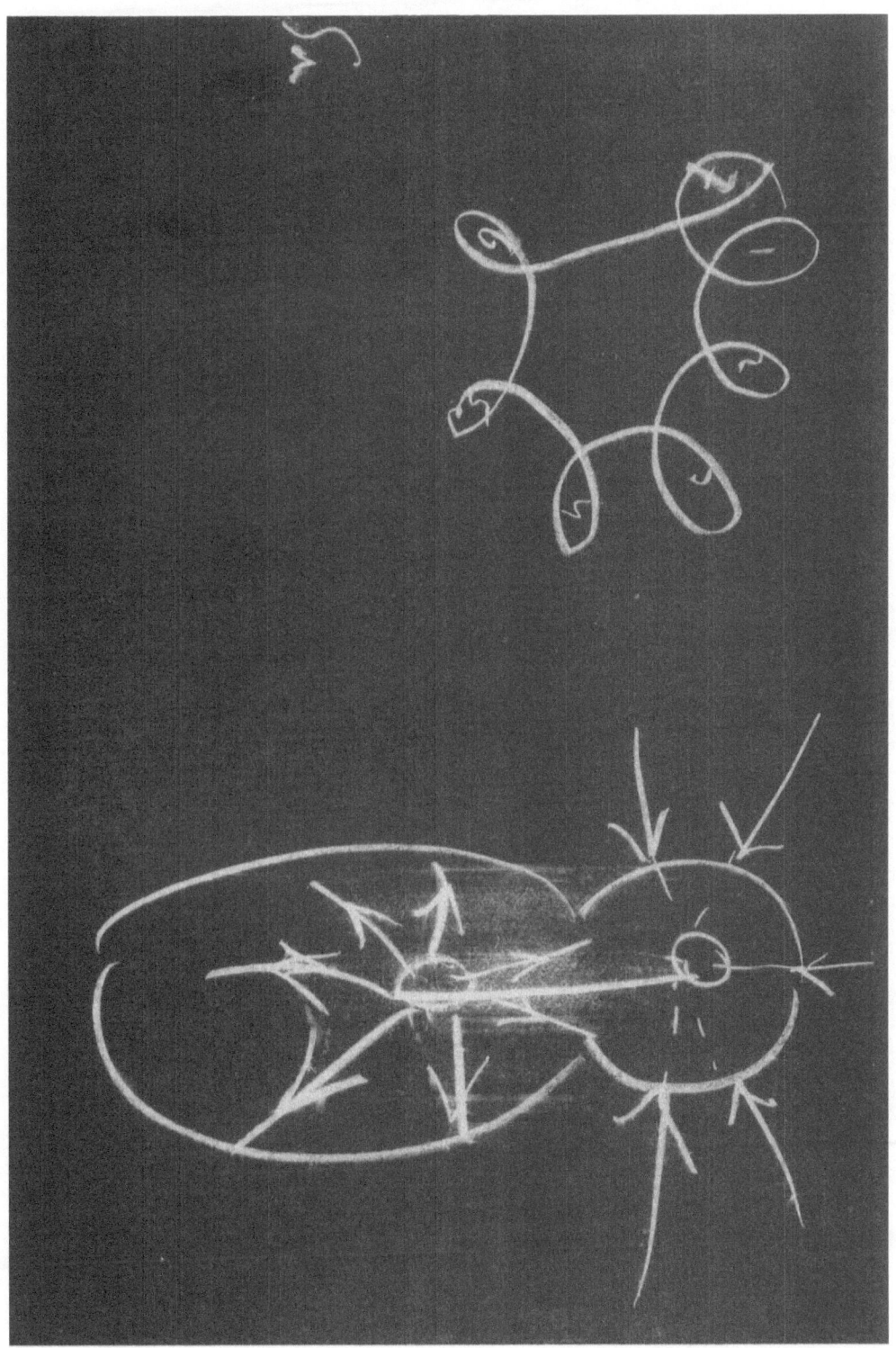

63

Plate 15

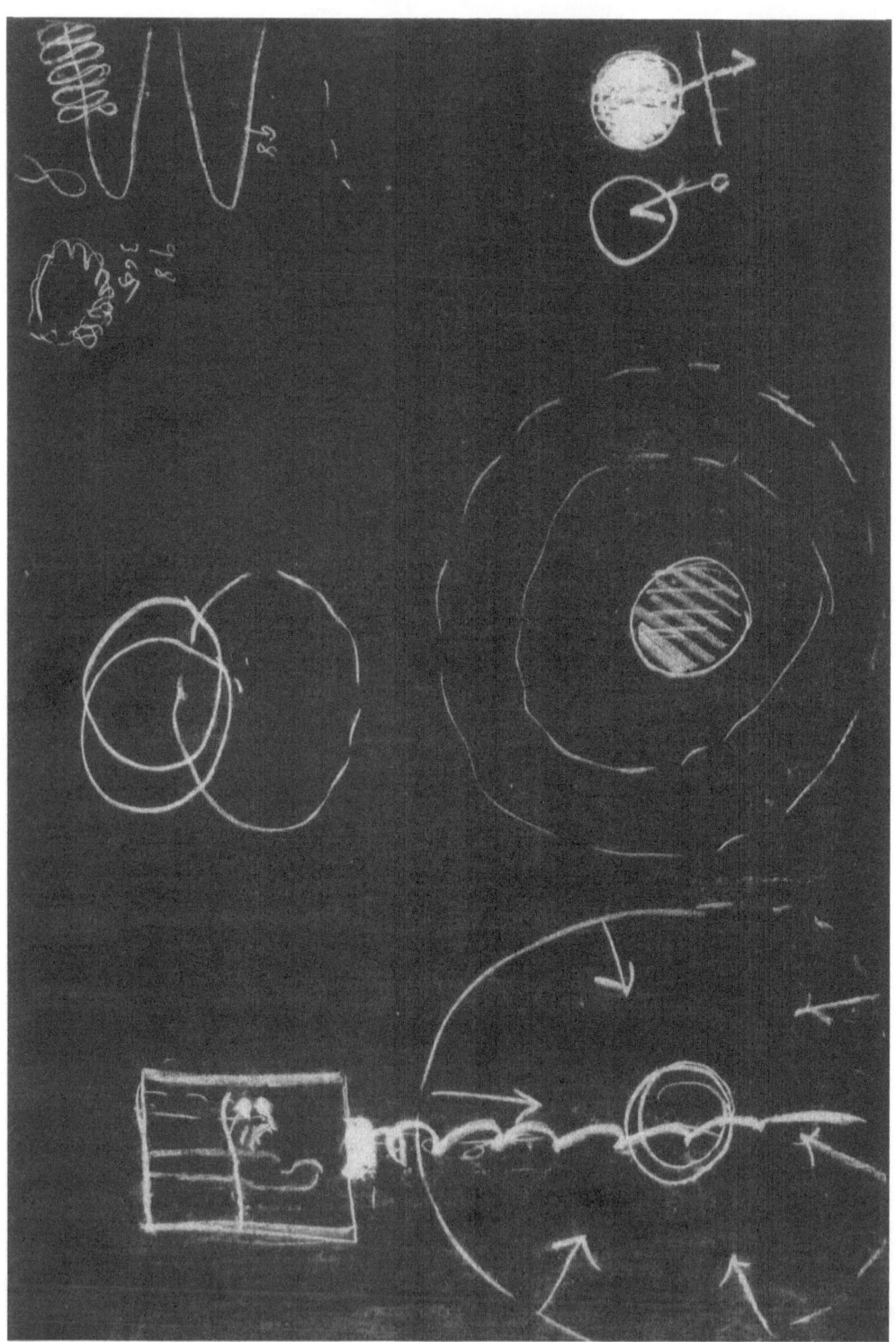

Plate 16

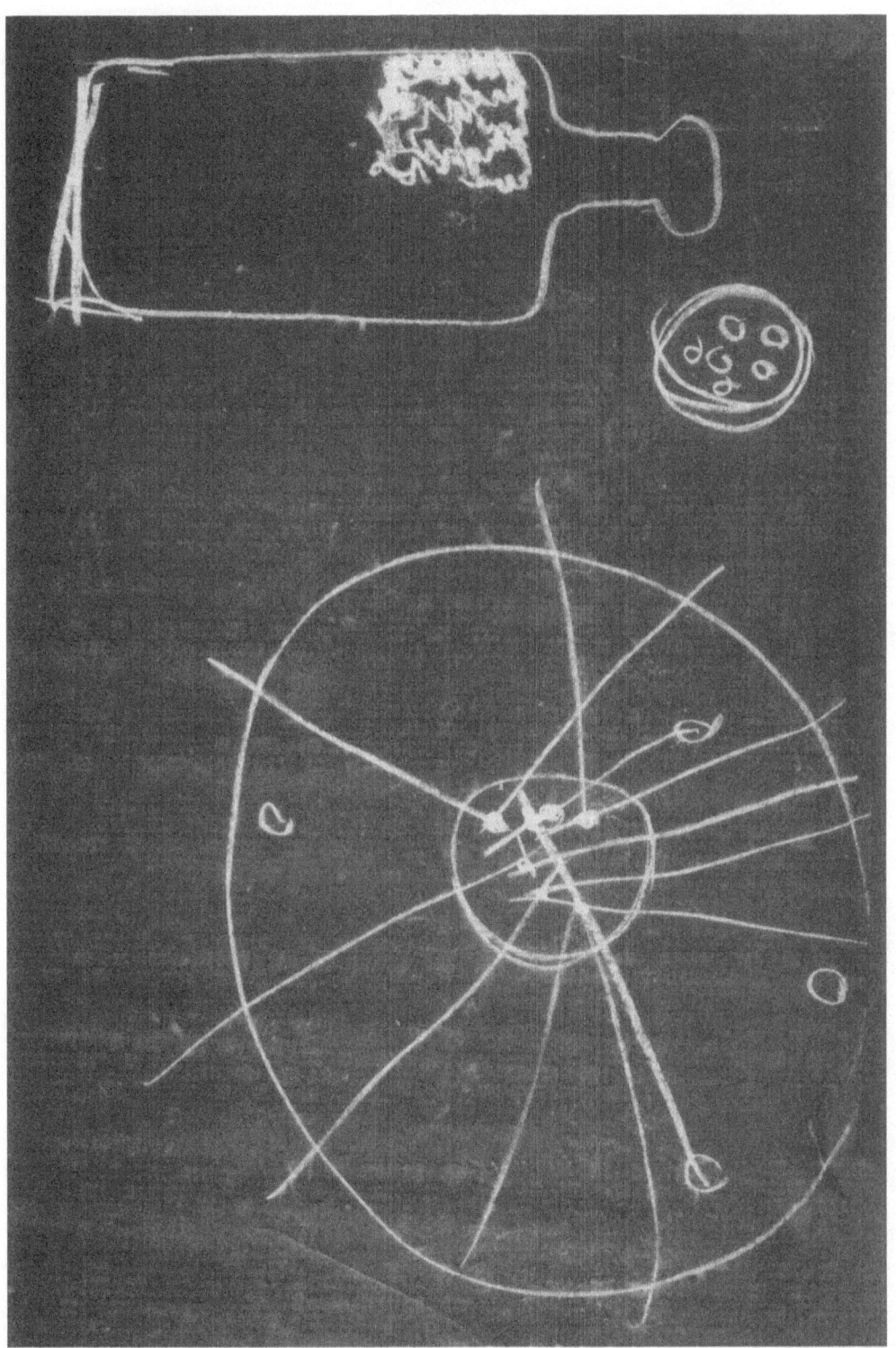

Plate 17

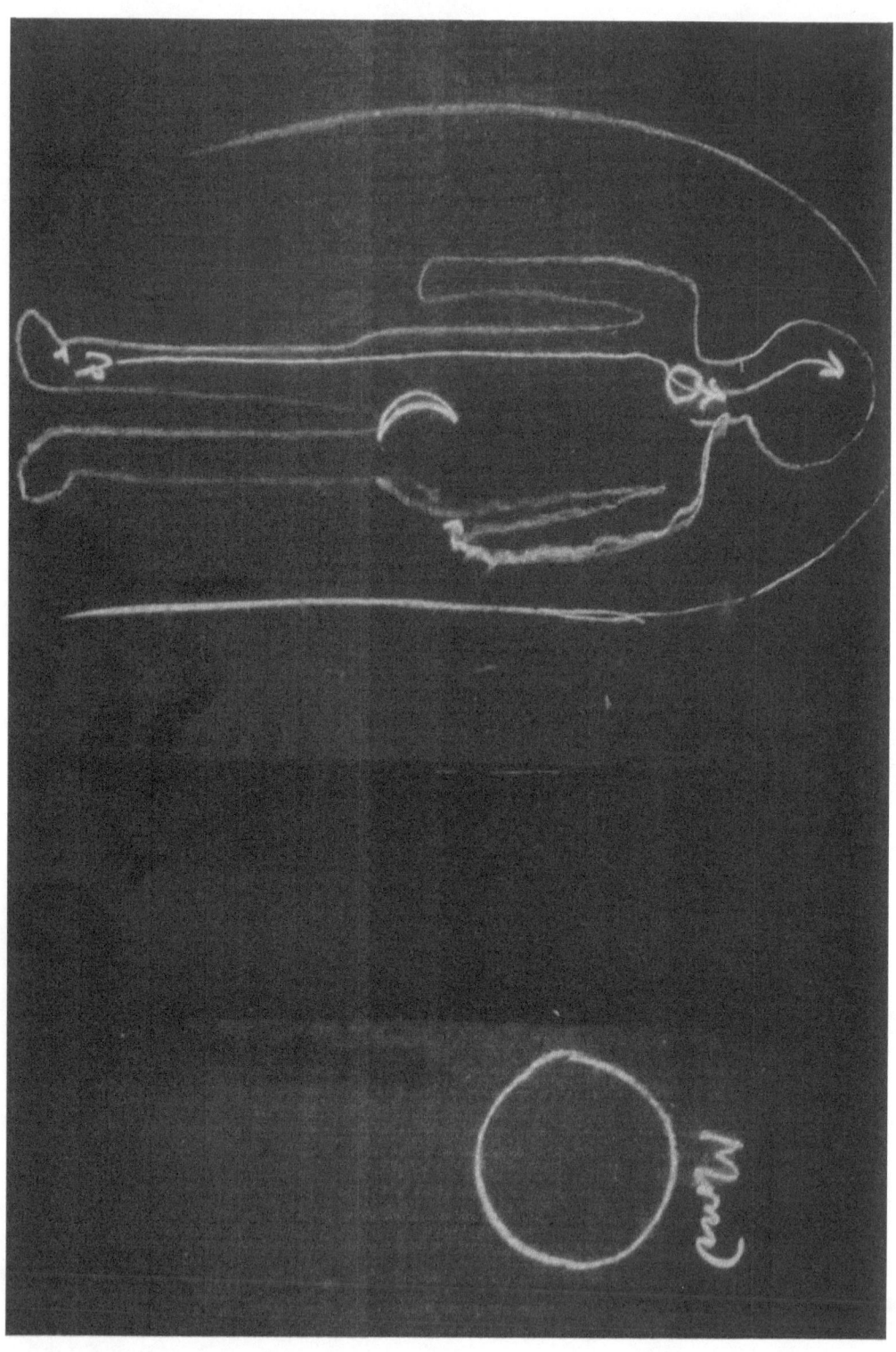

Plate 18

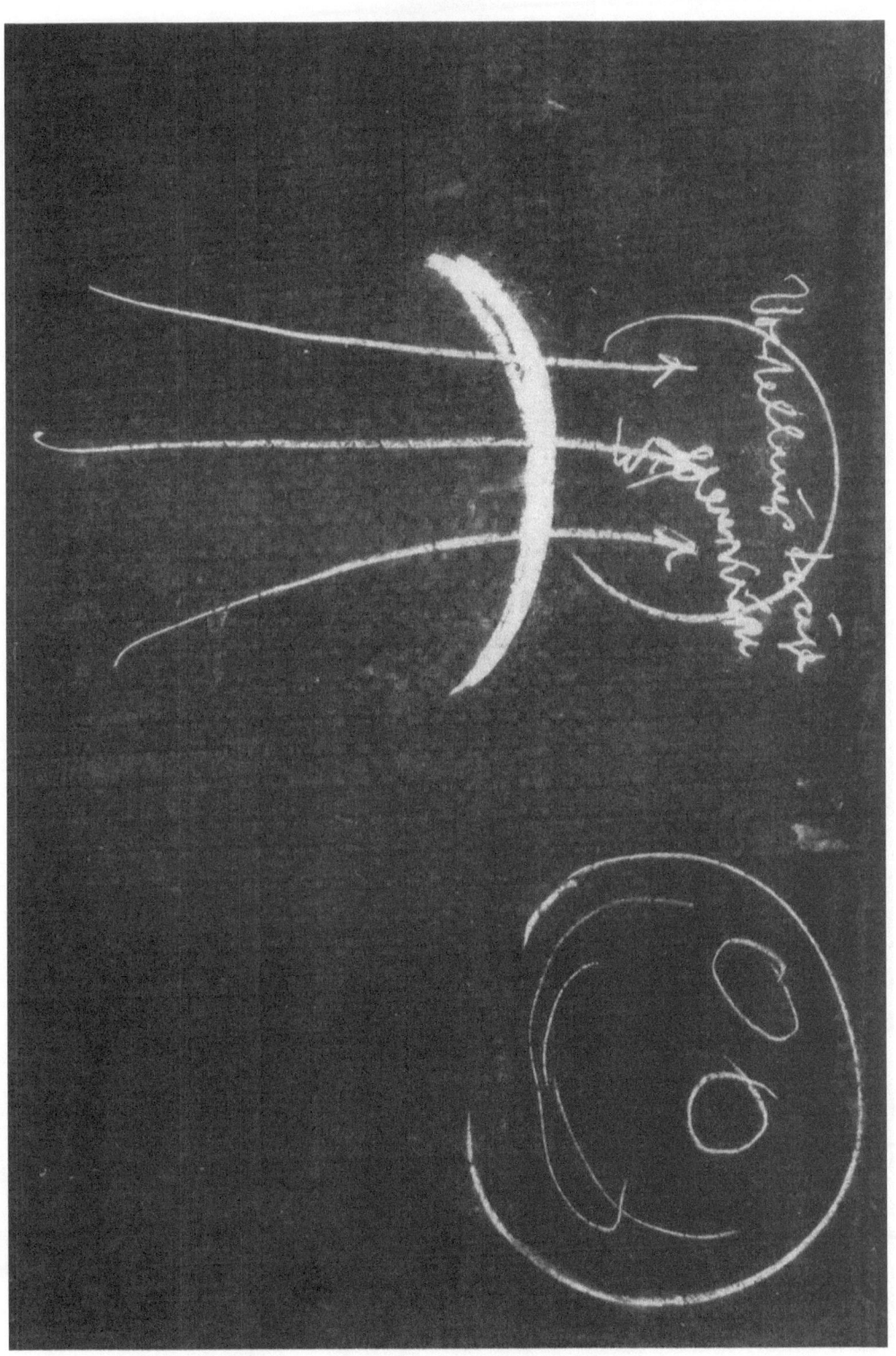

Plate 19

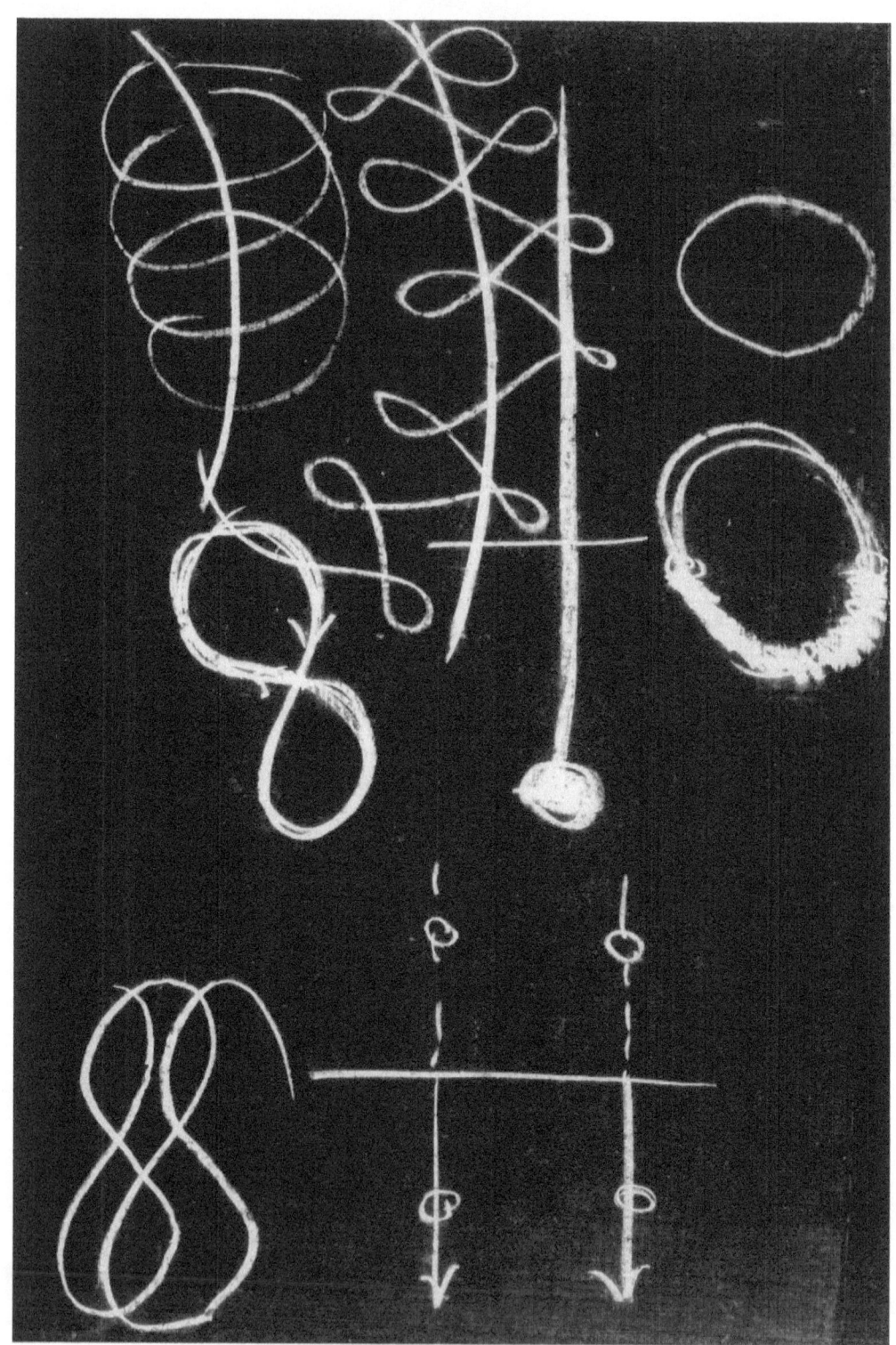

Plate 20

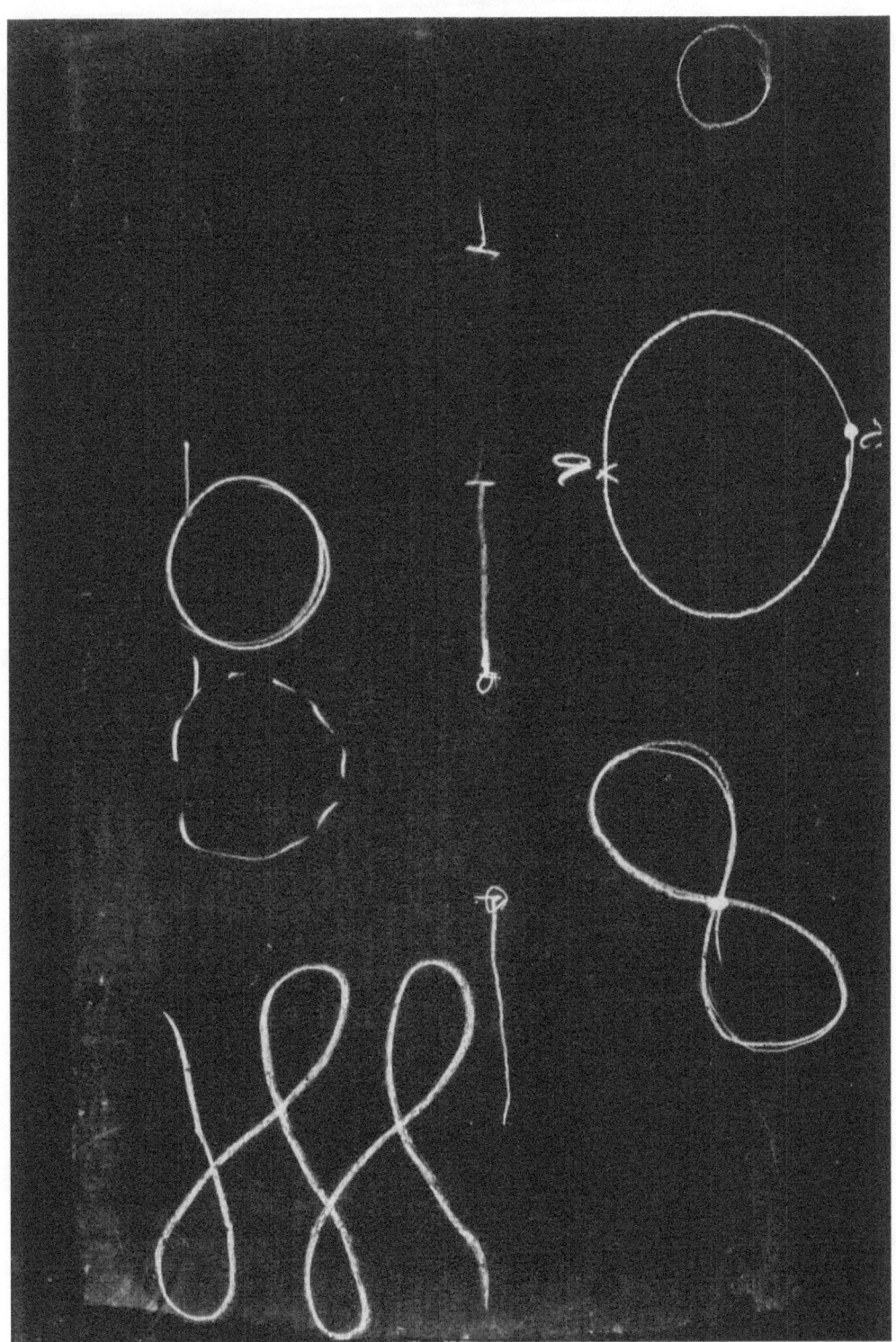

Plate 21

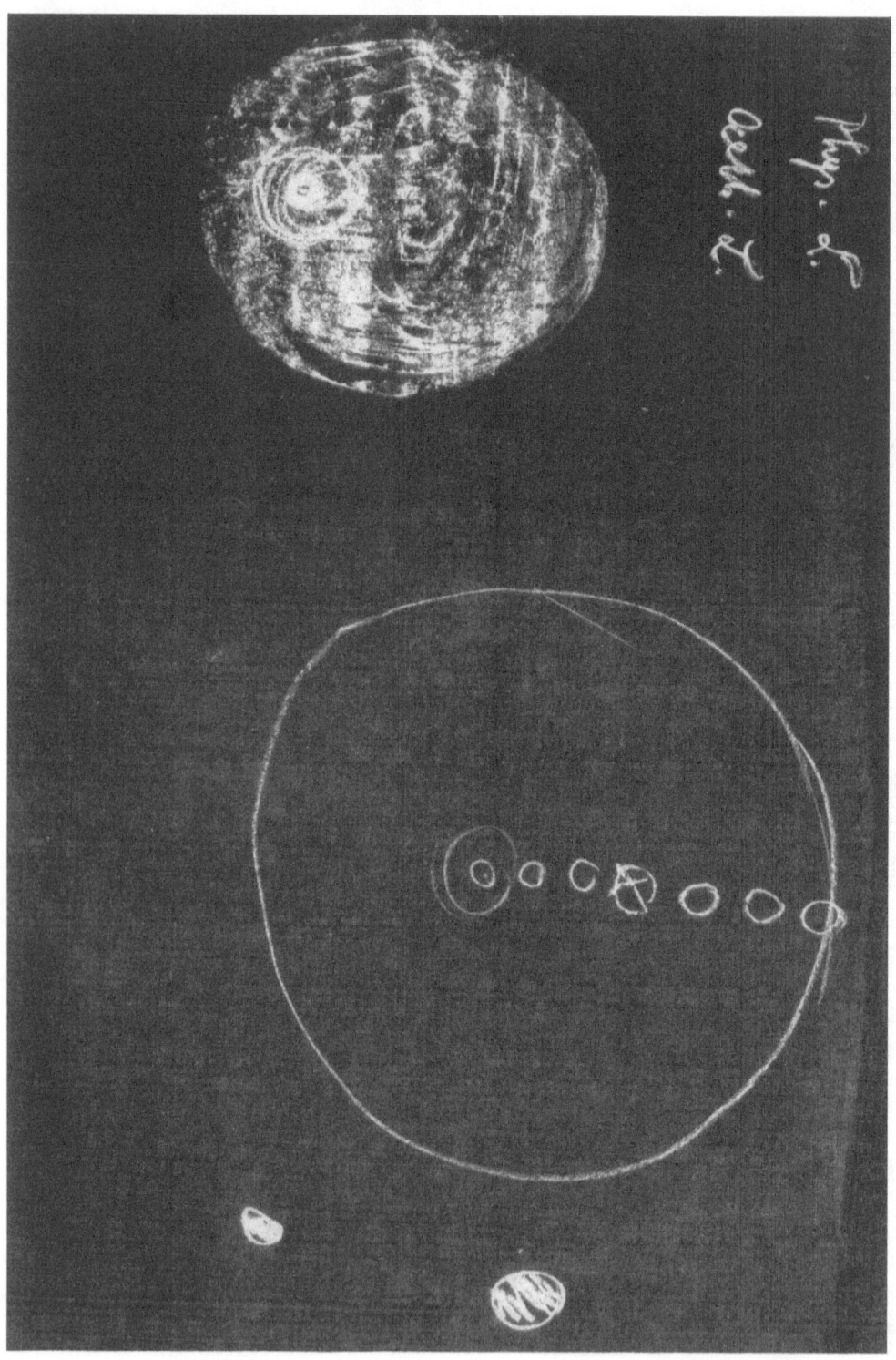

Plate 22

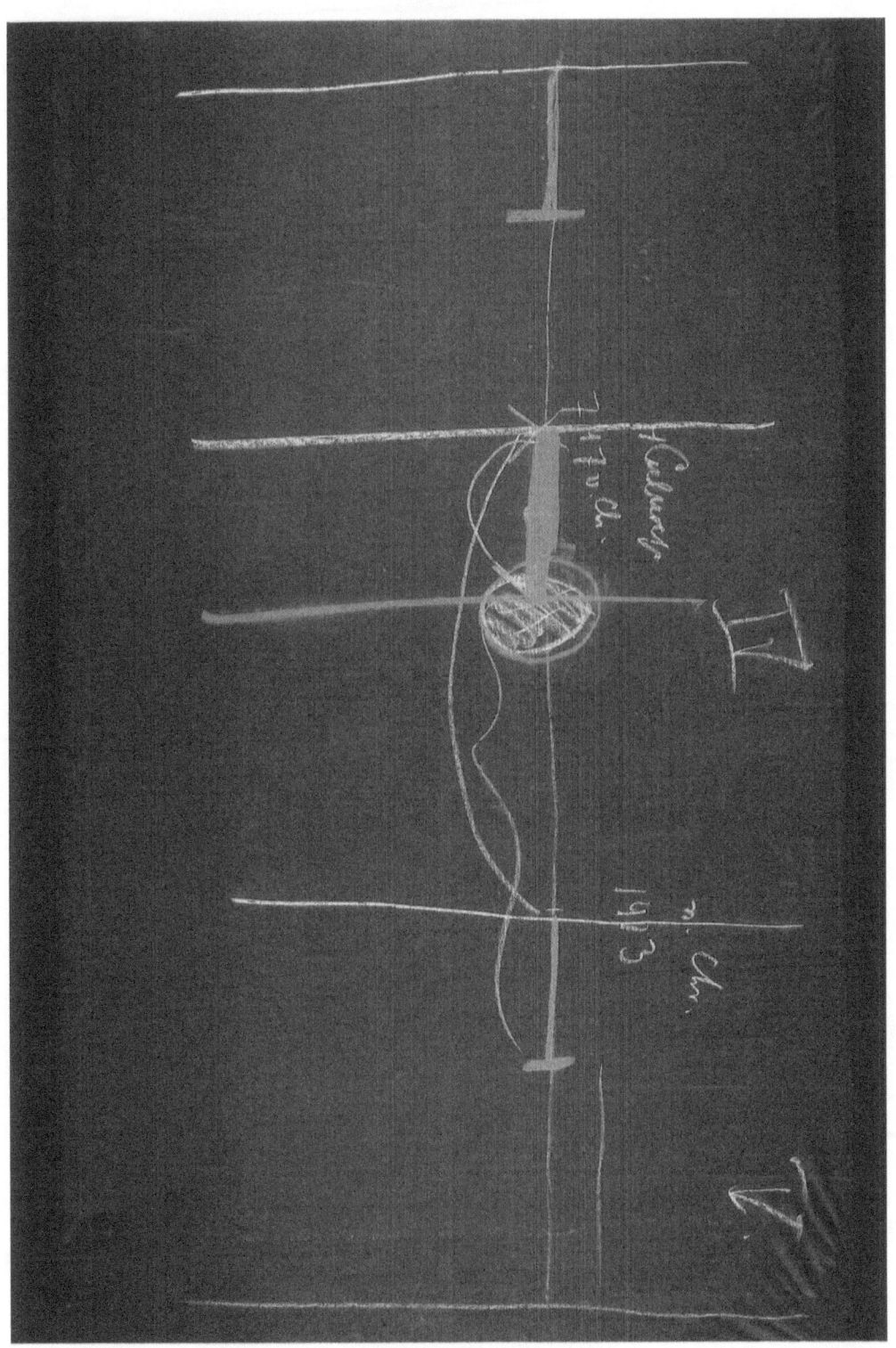

71

Plate 23

Plate 24

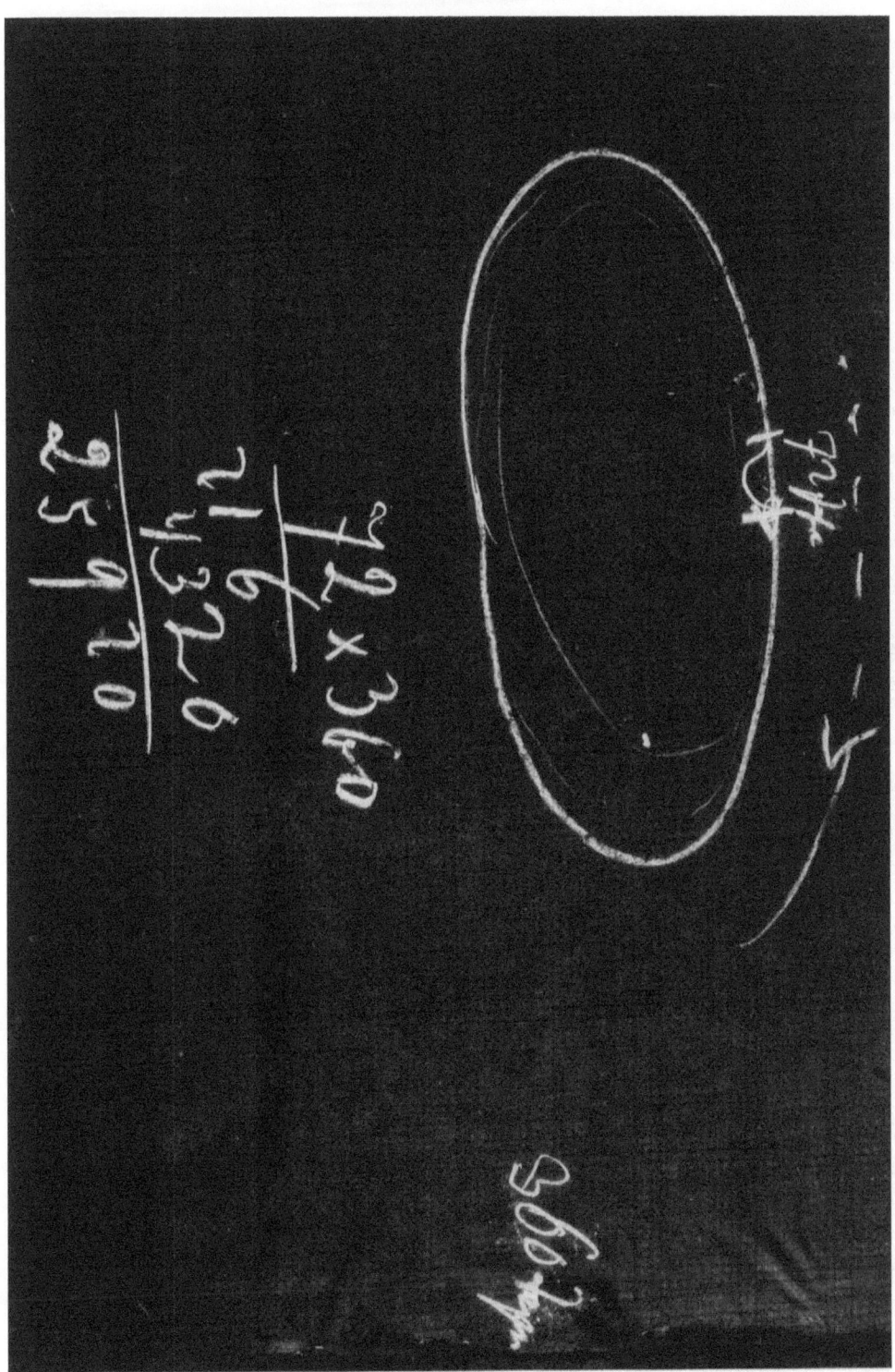

Plate 25

Plate 26

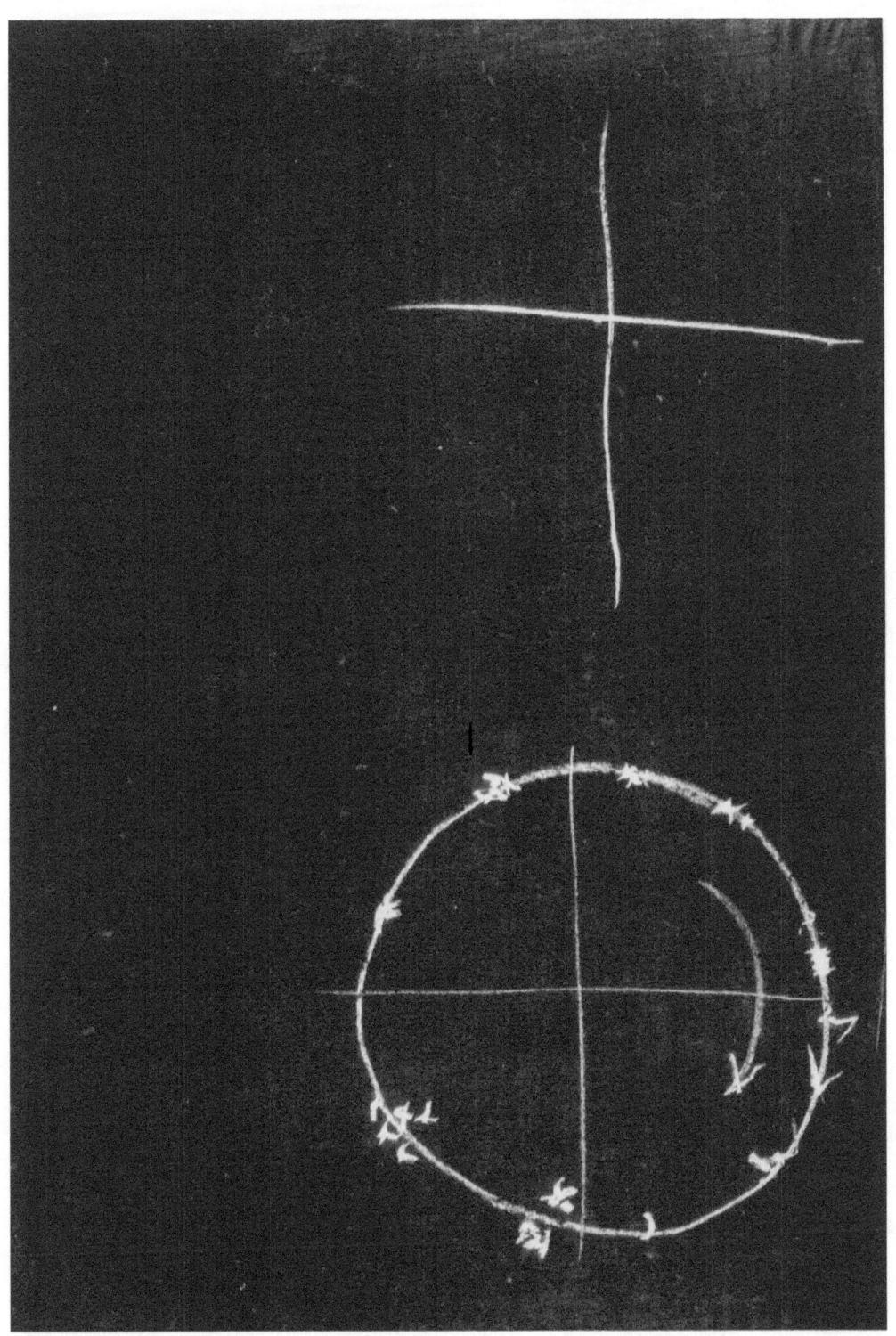

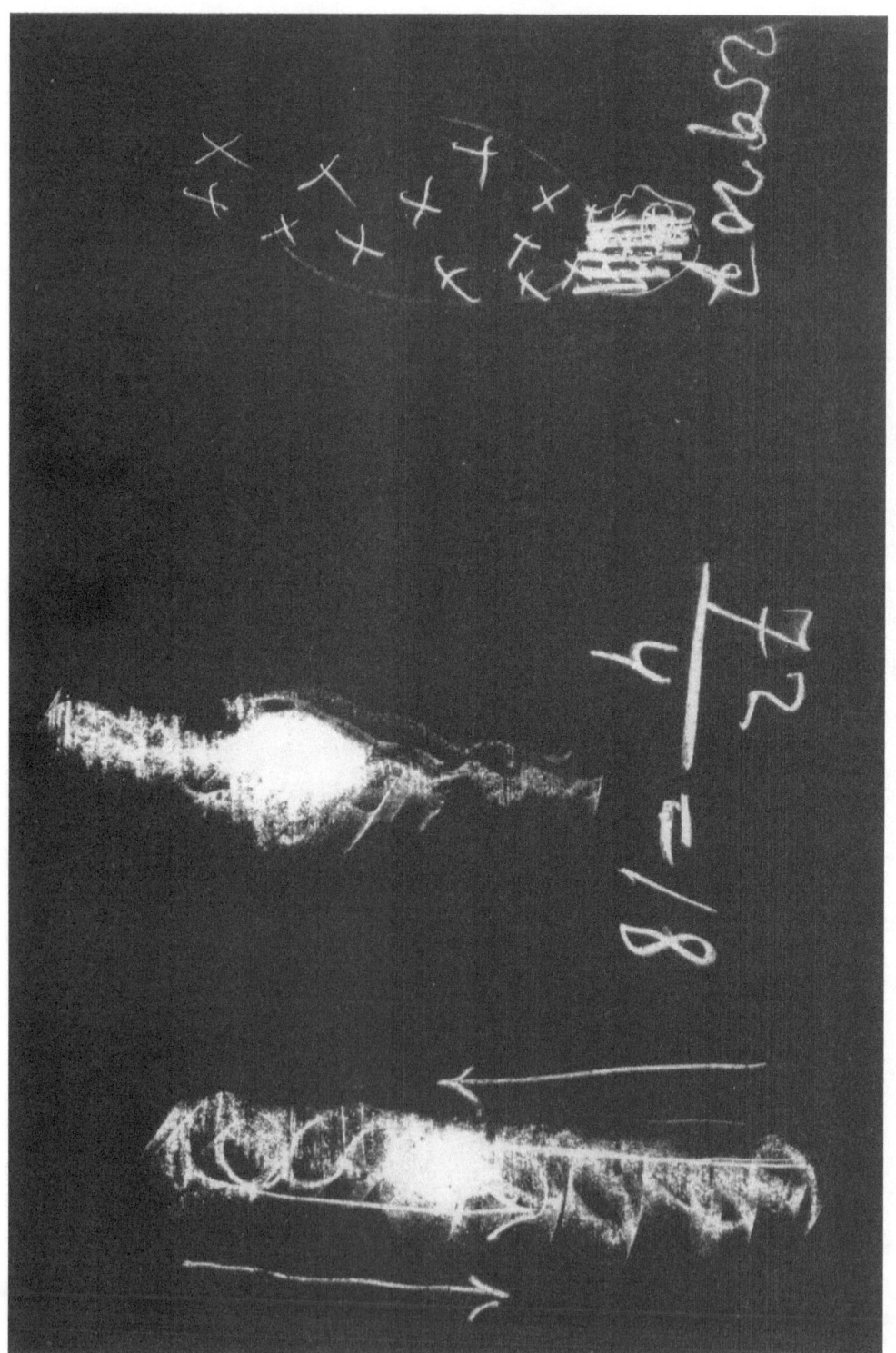

Plate 28

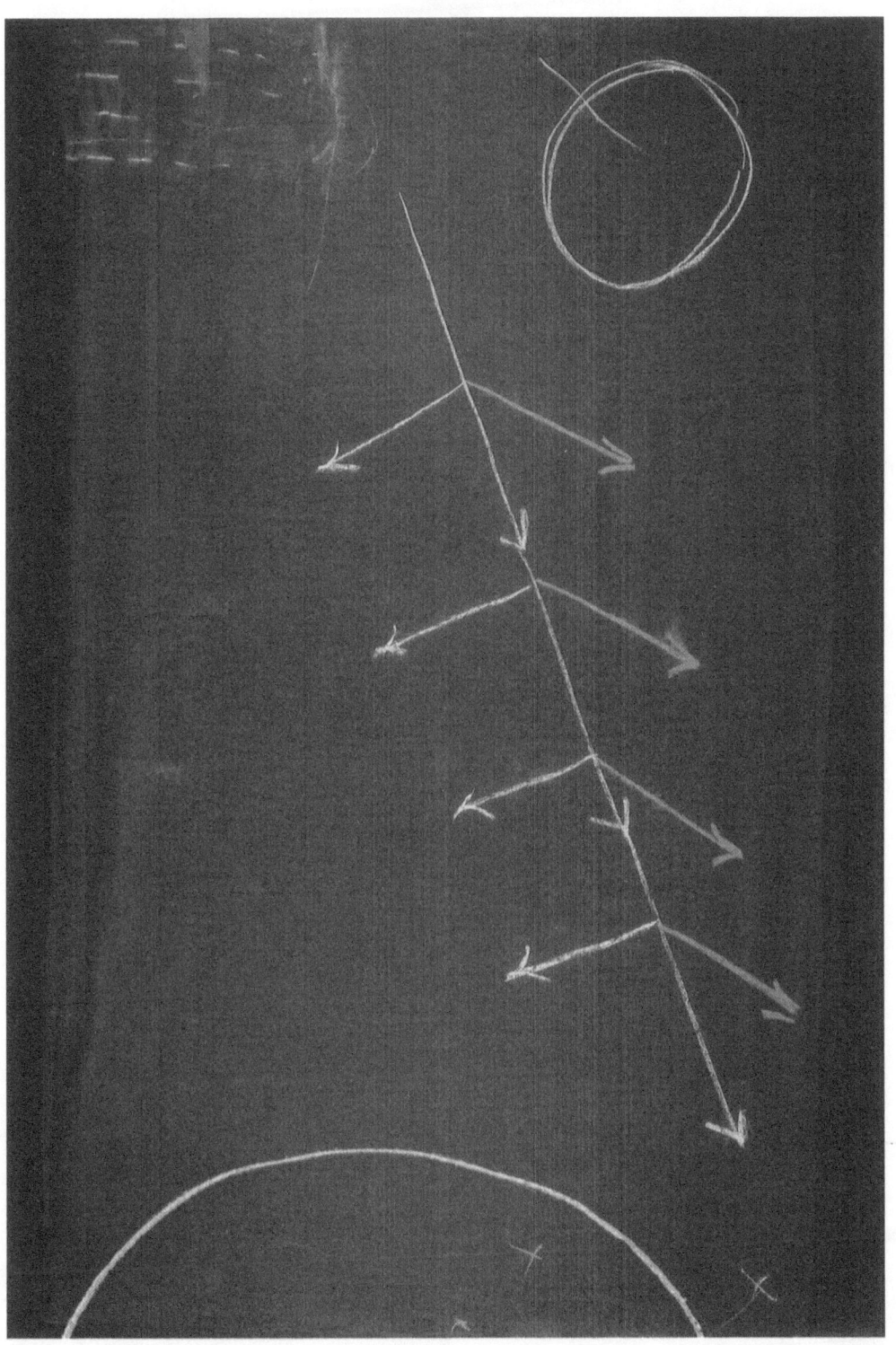

Plate 29

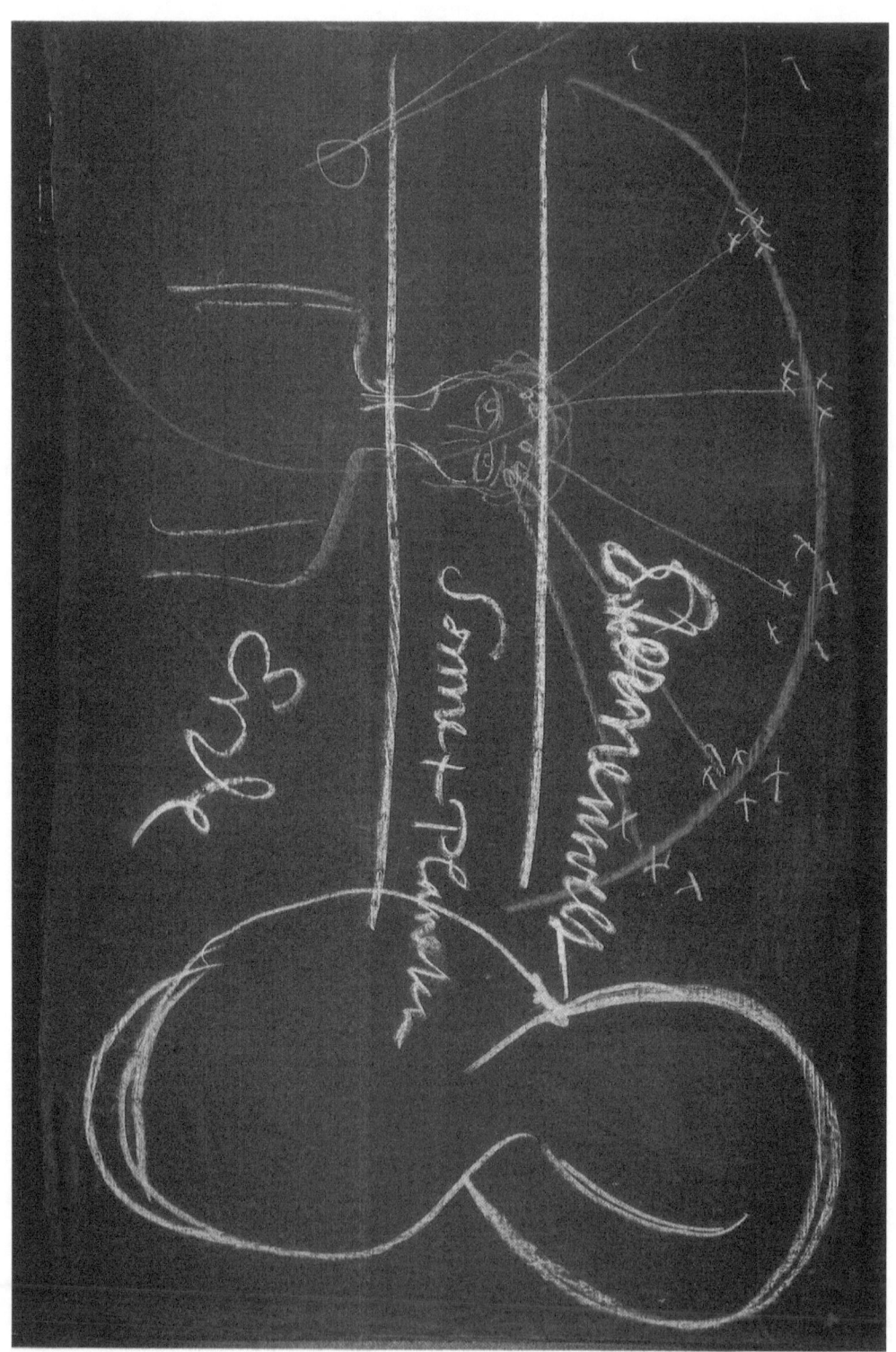

Plate 30

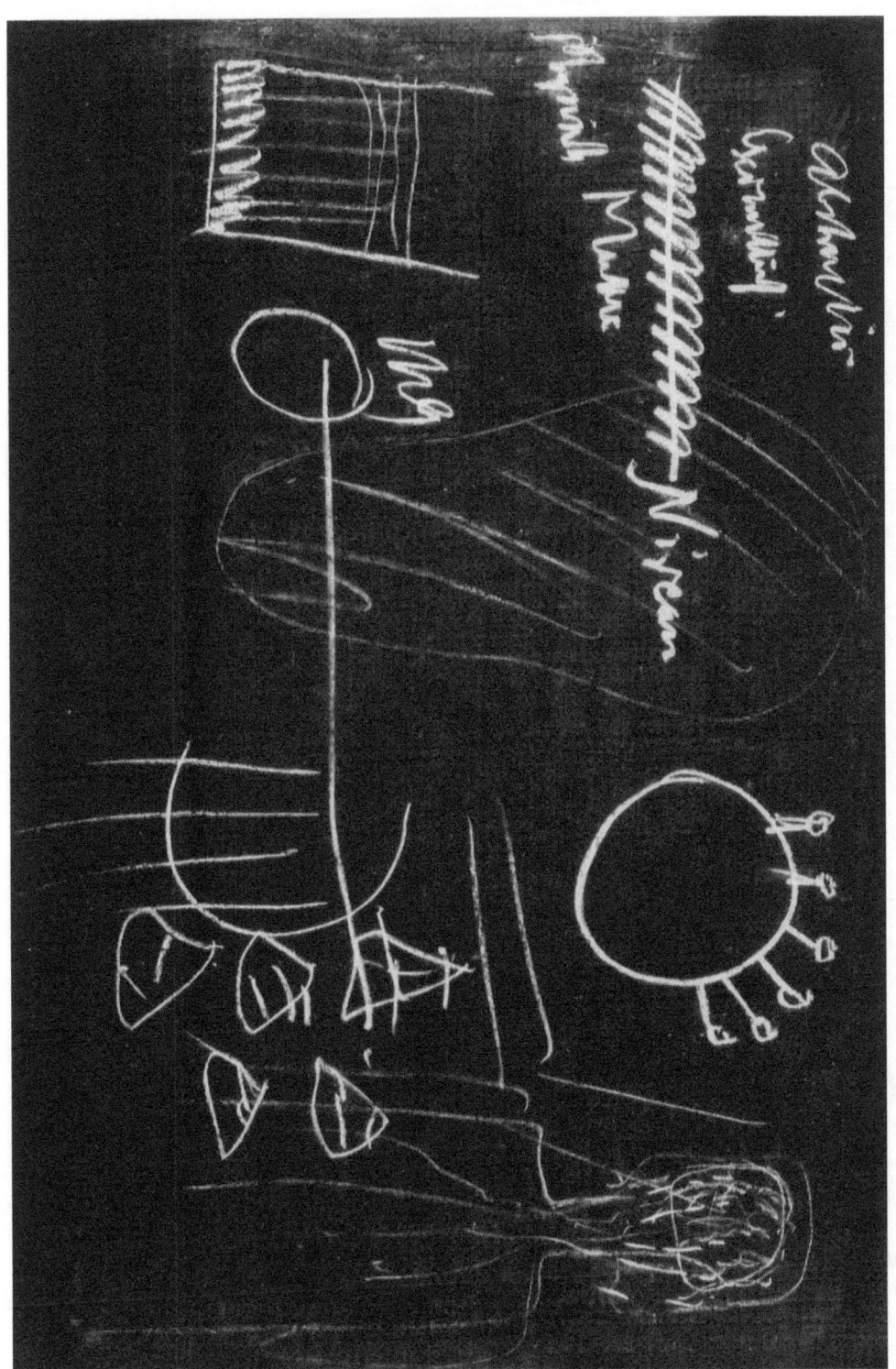

Plate 31

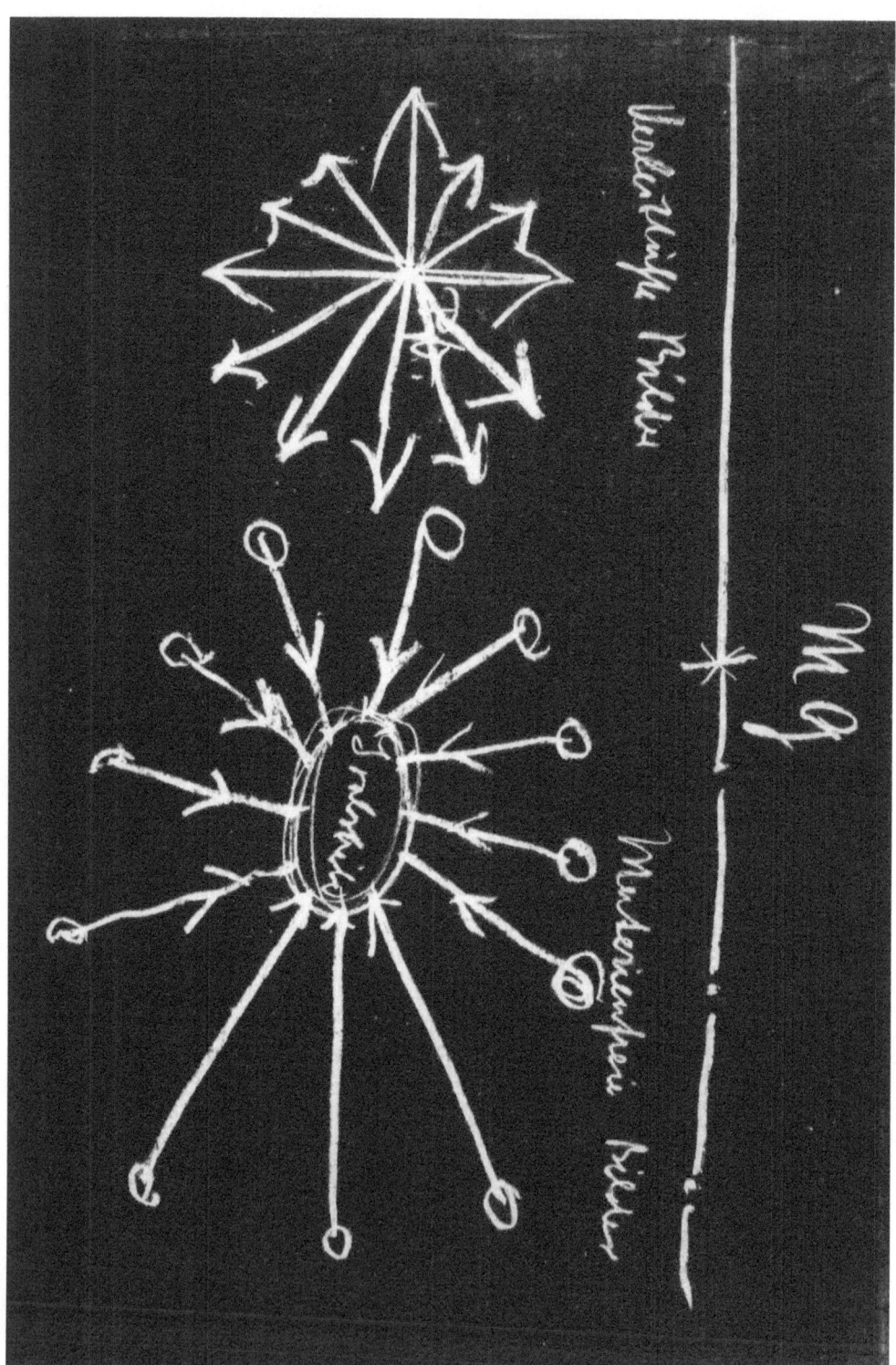

Notes

Arabic numerals = page numbers

t = top; m = middle; b = bottom

boldface = suggested emendation

2t **Kant** Immanuel Kant (1724-1804) was surely the most important modern philosopher, a great revolutionary who changed philosophy utterly and irrevocably. Hans-Georg Gadamer went so far as to argue that all subsequent philosophy has been little more than a series of attempts to rewrite Kant's *Third Critique*. He was the founder and the inspiration of both German Idealism and German philosophical Romanticism, and the German philosophers within those movements who influenced Steiner most directly (Goethe, Fichte, Schiller, and Hegel) were all profoundly indebted to Kant while simultaneously criticizing what they saw as shortcomings in his philosophy. Fichte and Schiller both described their philosophical work as attempts to rewrite the letter of Kant in the spirit of Kant. As a philosopher, Steiner was also both indebted to Kant and at the same time critical of him, although his most trenchant criticisms are usually directed not at Kant's actual positions, but rather the reductive neo-Kantianism of Steiner's contemporaries. For more on Steiner's difficult relationship to Kant, see Frederick Amrine, "Rudolf Steiner as a Philosopher," *Research Bulletin* [of the Research Institute for Waldorf Education], 19, No. 1 (2014), pp. 7-18.

7b **Galileo** Galileo Galilei (1564-1642), the Italian physicist, astronomer, philosopher, and mathematician "who made fundamental contributions to the sciences of motion, astronomy, and strength of materials and to the development of the scientific method. His formulation of (circular) inertia, the law of falling bodies, and parabolic trajectories marked the beginning of a fundamental change in the study of motion. His insistence that the book of nature was written in the language of mathematics changed natural philosophy from a verbal, qualitative account to a mathematical one in which experimentation became a recognized method for discovering the facts of nature. Finally, his discoveries with the telescope revolutionized astronomy and paved the way for the acceptance of the Copernican heliocentric system, but his advocacy of that system eventually resulted" in his being brought before the Inquisition. *(Britannica)*

7b **Copernicus** Nicolaus Copernicus (1473-1543) was the first astronomer to publish a comprehensive and persuasive heliocentric account of the universe, *On the Revolutions of the Heavenly Spheres* (1543).

11m **a representation of the ego** See Appendix 1.

13t **I was able to show the doctors** In the cycle *Introducing Anthroposophical Medicine* (see previous note in the book).

15b **but the forces, the *formative forces* which condition him** See Appendix 2.

20m **through the methods of spiritual science, to an imaginative conception of the universe** Steiner refers constantly to Imagination, Inspiration, and Intuition. The most systematic treatment is arguably GA 12; *The Stages of Higher Knowledge: Imagination, Inspiration, Intuition* (Great Barrington, MA: SteinerBooks, 2009). Like Wordsworth, Steiner saw Imagination as more than reason, "reason in its most exalted mood." A good way to understand Imagination, Inspiration, and Intuition is to recall, and then expand, Plato's Allegory of the Cave, in which everyday thoughts are revealed to be but the shadows of a higher, more active and intuitive thinking activity. Steiner followed the German Idealists and the Romantics in calling this mode of intuitive thinking Imagination. By the same token, Steiner argued that there are higher modes of cognition of which everyday feeling and willing are but the shadows of Inspiration and Intuition respectively.

21m **in my book *How to Know Higher Worlds*** CW 10; *How to Know Higher Worlds: A Modern Path of Initiation*, trans. Christopher Bamford, Classics in Anthroposophy (Great Barrington, MA: Anthroposophic Press, 1994).

21b **since the beginning of this fifth post-Atlantean epoch** "The Fifth Post-Atlantean Cultural Epoch" is Rudolf Steiner's theosophical term for the current historical epoch, which began ca. 1413 – i.e., with the Renaissance. Steiner correlated these historical epochs with the precession of the equinox through the signs of the zodiac, with each epoch lasting 2,160 years. He sees the Fifth as lasting until approximately 3573 CE.

22b   After yesterday's lecture Dr. Stein has taken the trouble

Walter Johannes Stein (1891-1957) was one of the founding teachers at the original Waldorf School in Stuttgart. Among other things, he went on the write an important history of the Grail (in English as *The Ninth Century and the Holy Grail* [Temple Lodge Publishing, 2001]). He was an important figure in the early history of the Anthroposophical Society.

25m   This was brought home to me in the question-yearnings arising out of the last lecture   *Frage-Sehnsuchten*, a strong neologism

48b **ether**   This is not Steiner's "etheric," but rather the refined medium imagined suffusing the universe by late 19th-century physics.

48b **theosophy**   Steiner's constant complaint against his erstwhile theosophical colleagues was that they were actually just a different kind of materialists. He accused them of imagining spirit as simply more attenuated matter.

55b **Schelling** Friedrich Wilhelm Joseph Schelling (1775-1854), the youngest of the four main philosophers in the movement conventionally known as German Idealism, the others being Immanuel Kant (1724-1804), Johann Gottlieb Fichte (1762-1814), and Georg Wilhelm Friedrich Hegel (1770-1831). Like Hegel, Steiner was often critical of Schelling, and for the same reasons, chiefly his desire to investigate nature by employing abstract metaphysical principles apart from, or even prior to, empirical experience. But Steiner clearly admired much in Schelling, especially his later works, which scholars are only just beginning to appreciate today.

56t **Goethe** Johann Wolfgang von Goethe (1749-1832), author of *Faust* and *Wilhelm Meister*, is widely regarded as the greatest German poet. He was also an extraordinary scientist: see his *Metamorphosis of Plants* (1790) and his *Theory of Color* (1810) especially.

64b **Bischoff** Th. L. W. Bischoff (1809-1882) *Die Großhirnwindung bei den Menschen*, München, 1968; *Studium und Ausübung der Medizin durch Frauen*, München, 1872.

70t **formative forces** See Appendix 2.

71t **How could we see the sun?** "Were the eye not sunlike / It never could gaze upon the sun" ["Wär' nicht das Auge sonnenhaft, / Die Sonne könnt' es nie erblicken."] Plotinus (205-270), *Enneads* I.6.9: "No eye ever saw the sun without becoming sun-like ..." (Plotinus, *Porphyry on Plotinus; Ennead I*, Loeb Classical Library [Cambridge, Massachusetts: Harvard University Press, 1966], p. 261)

73b **Foucault** J. B. L. Foucault (1819-1868). The experiment with the pendulum was performed in 1851 in the Pantheon.

74m **Einstein** Albert Einstein (1879-1955) was a German-born theoretical physicist. In 1905 and 1915 he published the Special and General Theories of Relativity respectively. Einstein is rightly admired for his genius, which Steiner acknowledges elsewhere.

86m ***Occult Science – An Outline*** CW 13; *An Outline of Esoteric Science*, trans. Catherine E. Creeger (Great Barrington, MA: SteinerBooks, 1997). An otherwise excellent earlier translation by George Adams bore the unfortunate title *Occult Science*. At the time of writing, Steiner was the head of the Theosophical Society in Germany, and the word *Geheimwissenschaft* in his title was meant to echo Blavatsky's tome, *The Secret Doctrine*. GA 13 is now considered one of the four "basic books" of anthroposophy.

87m **here I would like to proceed from something** In GA 301; *The Renewal of Education* (Great Barrington, Massachusetts: Anthroposophic Press, 2001).

87b **he mentions so-called subjective colours** See Johann Wolfgang von Goethe, *Scientific Studies*, vol. 12 of the Suhrkamp Edition [in English], ed. and trans. Douglas Miller (New York: Suhrkamp Publishers, 1988), pp. 168-185.

92b **recursive oscillation** *Zurückschwingungen*, a neologism.

94t **an externalization of the etheric liver** *Exsudat*, an unusual medical term

98m **Karl Gegenbaur** Carl Gegenbaur (1826-1903), anatomist. The relevant publications are "Über die Kopfnerven von Hexanchus und ihr Verhältnis zur 'Wirbeltheorie' des Schädels, *"Jenaer Zeitschrift f. Naturwiss.,* 6 (1871); "Das Kopfskelett der Selachier, ein Beitrag zur Erkenntnis der Genese des Kopfskeletts der Wirbeltiere. Untersuchung zur vergl. Anatomie der Wirbeltiere," 3. Heft, Leipzig 1872. *[B]* "The work by which Gegenbaur is best known is his *Grundriss der vergleichenden Anatomie* (Leipzig, 1874; 2nd edition, 1878), translated into English by W. F. Jeffrey Bell (as *Elements of Comparative*

Anatomy, 1878), with additions by E. Ray Lankester. While recognizing the importance of comparative embryology in the study of descent, Gegenbaur laid stress on the higher value of comparative anatomy as the basis of the study of homologies, i.e. of the relations between corresponding parts in different animals, as, for example, the arm of man, with the foreleg of a horse, and with the wing of a fowl. A distinctive piece of work was effected by him in 1871 in supplementing the evidence adduced by Huxley in refutation of the skull-vertebrae theory: the theory of the origin of the skull from expanded vertebrae, which, formulated independently by Goethe and Oken, had been championed by Owen. Huxley demonstrated that the skull is built up of cartilaginous pieces; Gegenbaur showed that in the lowest (gristly) fishes, where hints of the original vertebrae might be most expected, the skull is an unsegmented gristly brain-box, and that in higher forms, the vertebral nature of the skull cannot be maintained, since many of the bones, notably those along the top of the skull, arise in the skin."
[*Encyclopedia Britannica*, 11th edition, quoted in *Wikipedia*]

101t **Egypto-Chaldaean culture**   Steiner's theosophical term for a long epoch he sees as having ended ca. 747 BCE. He also calls it the third post-Atlantean epoch or age. The implication of the term is that the Egyptian, Sumerian and Babylonian cultures were in the vanguard of human cultural evolution during that period – quite a conventional notion after all, despite the idiosyncrasy of the label. Because it has become a standard anthroposophical term, "Chaldean" stands here rather that Sumerian and/or Babylonian, which would be more conventional in English.

106m   **kama-manas**   This theosophical term is roughly equivalent to the astral body. "Kama" is the lower astral body, filled with passions and desires, and the basis of personality. "Manas" is the higher astral nature, which turns away from attachment to the body towards the spirit; it is roughly equivalent to the consciousness soul.

111t   **there is a box**   This passage is contained in Einstein's treatise on Special and General Relativity (Braunschweig, 1917, pp. 45-48.) The thought experiment actually employs a stone and wood, rather than a feather. In 1920, the Theory of Relativity was widely discussed because it had successfully predicted an eclipse of the Sun. One of Steiner's fullest discussions of Einstein's Theory of Relativity is in his book CW 18; *The Riddles of Philosophy* (1973; Great Barrington, Massachusetts: SteinerBooks, 2009). There he proposes, intriguingly, that the Theory of Relativity will open the door for many to anthroposophy.

1123t **the General Meeting** On April 25th the 7th General Meeting of the "Goetheanum League" took place.

116b **of establishing such an institute** In 1920, a research institute was formed in the context of Steiner's larger social impulse, "The Coming Day." It was devoted principally to chemistry and physics, but it also had a biological section, which was moved to Dornach a few years later.

118m **eurythmy performances** See Appendix 3.

121t **Ms Molt** Berta Molt (1876-1939), the wife of the founder of the first Waldorf School in Stuttgart, and a teacher at that school.

122t **Marxists** See Appendix 4.

123t **De Rochas** Eugène Auguste Albert de Rochas d'Aiglun (1837-1914) was a leading French parapsychologist, among other things.

123b **the event of Golgotha** The Mystery [or event] of Golgotha is Steiner's favored term for all that is more conventionally referred to as the Incarnation, the Passion, and the Resurrection. One of Steiner's earliest publications, his book *Christianity as Mystical Fact*, situates Christianity squarely within the context of ancient mystery religions (rather than mysticism, which the title might seem to imply). This important text is now volume 8 of Steiner's *Complete Works* in English, and the latest edition (New York: Anthroposophic Press, 1997) was both translated and introduced by a scholar and anthroposophist of great stature, Andrew Welburn. In the same spirit, Steiner also refers to "the Easter Mystery."

127b **to develop ideas alone** See Appendix 1.

129m **Schopenhauer's *The World as Will and Idea*** Arthur
Schopenhauer (1788-1860), a post-Kantian German philosopher known
especially for his profound pessimism. Although Schopenhauer's
magnum opus, *The World as Will and Idea* [*Die Welt als Wille und
Vorstellung*] was first published in 1819, it was only decades later that
Schopenhauer became widely read and influential.

129b **Eduard von Hartmann** See Appendix 5.

129b **soul-will** *Seelenwille*, a neologism

130t ***Philosophy of Spiritual Activity*** CW 4; *The Philosophy of
Freedom: The Basis for a Modern World Conception: Some results of
introspective observation following the methods of Natural Science*,
trans. Michael Wilson (London: Rudolf Steiner Press, 1964). An
abridged version edited and translated by Frederick Amrine, can be
found on Amazon as *The Essential Philosophy of Freedom* (Keryx,
2017). Alternative English translations are available under the titles *The
Philosophy of Spiritual Activity* and *Intuitive Thinking as a Spiritual
Path*. Although Steiner did not identify it explicitly as such, CW 4 has
come to be viewed as one of the four "basic books" of anthroposophy.

133t **Kepler** Kant pronounced the German mathematician and
astronomer Johannes Kepler (1571-1630) the most rigorous thinker who
ever lived. He is most famous for discovering and mathematically
modeling the elliptical motions of the planets, and for major
contributions toward the theory of gravitation eventually formalized by
Newton.

135t **Lucifer** See Appendix 6.

135t **often spoken in public lectures** E.g. in the lecture of April 26, 1920 in GA 334; *Social Issues: Meditative Thinking & the Threefold Social Order* (Great Barrington, Massachusetts: SteinerBooks, 1992).

140b **the epicycloid theory** In Ptolemaic astronomy the epicycle "was a geometric model used to explain the variations in speed and direction of the apparent motion of the Moon, Sun, and planets. In particular it explained the apparent retrograde motion of the five planets known at the time." [*Wikipedia*]

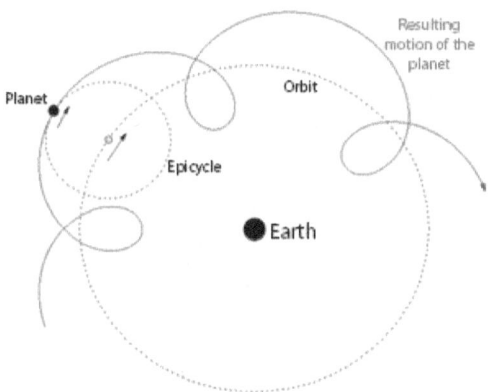

141b **Mercury's period of orbit** See the note to page 111t.

150m **Johannes Schlaf** Johannes Schlaf (1862-1941), an amateur astronomer, attempted in his works to refute the Copernican theory of the solar system.

1521t **All the planispheres** A planisphere is a revolving, two-dimensional star chart.

151t **Pythagoras** Pythagoras of Samos (c. 570 – c. 495 BCE) "was an ancient Ionian Greek philosopher … His political and religious teachings were well known in Magna Graecia and influenced the philosophies of Plato, Aristotle, and, through them, Western philosophy. … The teaching most securely identified with Pythagoras is *metempsychosis*, or the 'transmigration of souls', which holds that every soul is immortal and, upon death, enters into a new body. He may have also devised the doctrine of *musica universalis*, which holds that the planets move according to mathematical equations and thus resonate to produce an inaudible symphony of music." [*Wikipedia*]

151m **a *'counter-earth'*** *Gegenerde*, a neologism

153b **Lenin and Trotsky's message** Vladimir Lenin (1870-1924) "founder of the Russian Communist Party (Bolsheviks), inspirer and leader of the Bolshevik Revolution (1917), and the architect, builder, and first head (1917–24) of the Soviet state. He was the founder of the organization known as Comintern (Communist International) and the posthumous source of 'Leninism,' the doctrine codified and conjoined with Karl Marx's works by Lenin's successors to form Marxism-Leninism, which became the Communist worldview." [*Britannica*] Leon Trotsky (1879-1940) "was a Russian revolutionary, Marxist theorist, and Soviet politician whose particular strain of Marxist thought is known as Trotskyism. … Trotsky's ideas formed the basis of Trotskyism, a major school of Marxist thought that opposes the theories of Stalinism. He was written out of the history books under Stalin, and was one of the few Soviet political figures who was not rehabilitated by the government under Nikita Krushchev in the 1950s." [*Wikipedia*]

155t the Council held in Constantinople in 869 The Fourth Council of the Catholic Church, which decided among other things that humans had only a soul, and not also a spirit.

155t a Roman Jesuit, Father Secchi Angelo Secchi (1818-1878) was a professor at the Collegio Romano in Rome. He used the observatory erected in 1852 to investigate the physical qualities of the planets, the Moon, and the Sun. He was the first to divide the fixed stars into classes according to their spectra.

155b in Father Wasmann's study of ants Erich Wasmann (1859-1931), *Das Gesellschaftsleben der Ameisen [The Social Life of Ants]*, 1891.

157t I have already made mention of it E.g. in the cycle GA 138; *Intitation, Eternity and the Passing Moment* (Hudson, New York: Anthroposophic Press, 1980).

157m Pastor Kalthoff Albert Kalthoff, a Protestant theologian and for a period president of the German Monists' League.

158b the Kant-Laplace theory An important cosmogonic theory known as the "nebular hypothesis," and the ancestor of our contemporary "solar nebular disk model." It hypothesizes that the solar system was formed when a primordial nebula began to spin and formed planets through the resultant centrifugal force, and many found the explanation compelling because it seemed to explain the planets' circular and roughly coplanar orbits moving in the same direction as the Sun's rotation. It was first expounded fully by Immanuel Kant in his *Universal Natural History and Theory of the Heavens* [1755; multiple English editions available]. Pierre-Simon, Marquis de Laplace (1749-

1827), the great French mathematician and astronomer who extended the Newtonian paradigm in his five-volume *Celestial Mechanics* (1799-1825) and other writings, developed the same theory independently and published it in his *System of the World* [1796; many English translations available]. Steiner often derides this hypothesis, so often demonstrated in school classrooms, for failing to account the teacher-demiurge who sets the nebula in motion in the first place. (See e.g. the end of the lecture of January 18, 1921 in CW 323; *Interdisciplinary Astronomy* [(Amazon;) Keryx, 2019].)

159b **While doing so it also rotates around itself.** The latter rotation is uniform, but the Moon's rotation around the Earth is not. Thus libration results.

163t **biogenetic law** The biogenetic law states that *ontogeny recapitulates phylogeny*, i.e. that the history of the individual recapitulates the history of the race. It was widely held to be true in the nineteenth century.

165m **one of my recent public lectures** May 5, 1920, in CW 334.

166t **'Heaven and earth shall pass away, but my words shall not pass away'.** Matthew 24:35.

170m **The position could be determined by that of Sirius** In early Egyptian times the rising of the Nile corresponded to the helical rising of Sirius.

177t **into human souls** *Gemüter. Gemüt* is a word with no precise English equivalent that refers simultaneously to our thinking, feeling, sensibility, and temperament, all bundled up together.

177b **Adolf Harnack's *The Nature of Christianity*** Adolf von Harnack (1851-1930) was a prolific and highly influential German Protestant theologian and Church historian.

178m **Julius Robert Mayer** Julius Robert Mayer (1814-1878) was a German physicist, and a found.er of the field of thermodynamics. "He is best known for enunciating in 1841 one of the original statements of the conservation of energy or what is now known as one of the first versions of the first law of thermodynamics, namely that 'energy can be neither created nor destroyed'." [*Wikipedia*]

178b **Rümelin was the friend's name** Gustav Rümelin (1815-1888), a writer and politician. His recollections of Mayer are contained in his *Reden und Aufsätze* (3 vols., 1875-1894).

182b **Basilius Valentinus** Or Basil Valentine, traditionally identified as a Benedictine monk. Steiner confirms the traditional interpretation. The works published under this name were undoubtedly very adept, even in terms of modern chemistry, let alone alchemy. The *Twelve Keys* (1602) and *The Triumphant Chariot of Antinomy* (1604) are especially important. According to Steiner, Basil Valentine's works were actually compiled by his students, hidden, and then discovered and published by Johann Thölde.

183m **such concepts conceive** *wie es vorgestellt wird von dieser Vorstellungsart.* See Appendix 1.

183m **Huxley** Thomas Henry Huxley (1825-1895) was an English comparative anatomist, perhaps the finest of the later 19th century. His vigorous support of evolutionary theory earned him the nickname "Darwin's bulldog." Huxley coined the term "agnosticism." His famous debate with Samuel Wilberforce at Oxford was a key turning point in the public acceptance of evolutionary theory.

184t **luciferic powers** See Appendix 6.

186 **the Chaldaean Saros period** This period is almost exactly 18 years and 11 days, and is not identical with the periodicity of the Moon node of 18 years and 7 months.

186t **"God geometrizes, arithmetizes."** The quote is not actually to be found anywhere in Plato, but it is reported by Plutarch, and it is entirely in the spirit of his philopophy.

190t ***Christianity as Mystical Fact*** CW 8; *Christianity as Mystical Fact and the Mysteries of Antiquity*, trans. Andrew Welburn, ed. Christopher Bamford (Great Barrington: SteinerBooks, 2006).

193m **Wilhelm Wundt** Wilhelm Maximilian Wundt (1832-1920) was a pioneering German psychologist. Indeed, he coined the term "psychologist." He established a psychological laboratory, and in so doing established psychology as a discipline separate from both biology and philosophy. But Wundt's most important innovation was to conceive psychology as an experimental science. His work was wide-ranging and highly interdisciplinary; his method was largely – but not exclusively – materialist.

193m **non-philosopher** *Unphilosoph*

194t **Rubner in Berlin** Max Rubner (1854-1909), physiologist and hygienist.

194m **in my public lecture in Basel** Lecture of May 5, 1920 in CW 334.

197 **Novalis** Novalis was the pen name of Friederich von Hardenberg (1772-1801), a German Romantic poet, writer, and philosopher. The quote is from his *Apprentices at Sais*, Part 1.

198b **the ether** Here Steiner means the hypothetical medium of 19th-century physics, as opposed to what he usually means by this term.

201b **Einstein's box** See the note to page 111t.

201b **as thin as a sheet of paper** This is a consequence of the Lorentz-transformation as one approaches the speed of light.

202b **three lectures on the philosophy of Thomas Aquinas** GA 74; *The Redemption of Thinking* (Hodder & Stoughton, 1956; rpt. Anthroposophic Press, 2000).

202b **Augustinism**   Saint Augustine of Hippo (354-430) "was a Roman African, early Christian theologian and philosopher from Numidia whose writings influenced the development of Western Christianity and Western philosophy. He was the bishop of Hippo Regius in North Africa and is viewed as one of the most important Church Fathers in Western Christianity for his writings in the Patristic Era. Among his most important works are *The City of God, On Christian Doctrine* and *Confessions*." [*Wikipedia*]

205m **Eucken** Rudolph Eucken (1846-1926) was a German Idealist philosopher who won the Nobel Prize for Literature. "Distrusting abstract intellectualism and systematics, Eucken centred his philosophy upon actual human experience. He maintained that man is the meeting place of nature and spirit and that it is his duty and his privilege to overcome his nonspiritual nature by incessant active striving after the spiritual life. This pursuit, sometimes termed ethical activism, involves all of man's faculties but especially requires efforts of the will and intuition. A strident critic of naturalist philosophy, Eucken held that man's soul differentiated him from the rest of the natural world and that the soul could not be explained only by reference to natural processes." His main works were *Individual and Society* (1923); *Socialism: An Analysis,* 1921; *The Meaning and Value of Life,* 1909; and *Can We Still Be Christians?,* 1914. [*Britannica*]

205m **Bergson** Henri Bergson (1859-1941) was a great and highly influential philosopher who prized intuition above rationality, and was awarded the Nobel Prize in Literature. Bergson attempted to defend the possibility of human free will by radically redefining the notions of time, causality and thinking. His most important publications were *Time and Free Will* (1889), *Matter and Memory* (1896), *Creative Evolution* (1907), and *The Two Sources of Morality and Religion* (1932). In *The Riddles of Philosophy* (1973; [Great Barrington, Massachusetts:] SteinerBooks, 2009), Steiner applauds Bergson's epistemology while lamenting the flimsiness of his vitalistic theories of evolution.

210m **Descartes** "René Descartes (1596–1650) was a creative mathematician of the first order, an important scientific thinker, and an original metaphysician. During the course of his life, he was a mathematician first, a natural scientist or "natural philosopher" second, and a metaphysician third. In mathematics, he developed the techniques that made possible algebraic (or 'analytic') geometry. In natural philosophy, he can be credited with several specific achievements: co-framer of the sine law of refraction, developer of an important empirical account of the rainbow, and proposer of a naturalistic account of the formation of the earth and planets (a precursor to the nebular hypothesis). More importantly, he offered a new vision of the natural world that continues to shape our thought today: a world of matter possessing a few fundamental properties and interacting according to a few universal laws. This natural world included an immaterial mind that, in human beings, was directly related to the brain; in this way, Descartes formulated the modern version of the mind–body problem. In metaphysics, he provided arguments for the existence of God, to show that the essence of matter is extension, and that the essence of mind is thought. Descartes claimed early on to possess a special method, which was variously exhibited in mathematics, natural philosophy, and metaphysics, and which, in the latter part of his life, included, or was supplemented by, a method of doubt." [*Stanford Encyclopedia of Philosophy*]

216m in the chapter called "Sensory-Moral Effect of Color" See Goethe, *Scientific Studies*, pp. 278-297.

216b the three-volume work on Goethe by Father Baumgartner Alexander Baumgartner (1841-1910), author of *Goethe: Sein Leben und seine Werke [Goethe: His Life and His Works]*, 1879-1885. Writing as a Jesuit, his strongly Catholic perspective is apparent in all his works.

217m Weber's *Thirteen Lime Trees* Friedrich Wilhelm Weber (1813-1894), was a German doctor, politician and poet. "His reputation, however, was founded on his epic, *Dreizehnlinden* (1878). It enjoyed a wide circulation, and was arranged for the stage; he was nicknamed 'Dreizehnlinden-Weber' after it." [*Wikipedia*]

Appendices

Appendix 1

Representation

Die
Vorstellung

Translators' introductions or notes invariably comment on the difficulty of translating *vorstellen/Vorstellung* and a few other German philosophical and psychological terms such as *Geist*, *Anschauung*, and *Gemüt*. Tellingly, in the title of Schopenhauer's *magnum opus*, *Die Welt als Wille und Vorstellung*, has been translated three different ways: initially as "Representation," but then more recently as "Idea" and finally as "Presentation." Both Michael Wilson's translation of Steiner's *Philosophy of Freedom* and Owen Barfield's *The Case for Anthroposophy*[1] begin their discussions by noting that the standard translation of this Kantian philosophical term is "representation." But then Wilson goes on to argue (rightly) that this is too technical a term for most contexts, and that "representation" has other, distracting meanings outside of philosophy, which led him to translate *Vorstellung* and its variants as "mental picture." Barfield chooses "representation," which is understandable given that the context is Steiner's discussion of Brentano's neo-Kantian treatise. The reasons why this term is so very difficult to translate are (1) that it encompasses many different kinds of mental acts; (2) it refuses any sharp distinction between subjective and objective; (3) it can

[1] This is older, partial translation. For a newer translation, see Rudolf Steiner, *CW 21; On The Enigmas of the Soul,* trans. Frederick Amrine and Owen Barfield, commentary by Frederick Amrine ([Amazon:] Keryx 2017).

refer to a faculty, the activity of a faculty, or the result of that activity; (4) it often has a distinctly visual quality, but it can also refer to abstract concepts; and (5) it straddles the conventional divide between philosophy and psychology.

Appendix 2

The Etheric and the Astral Bodies

"Etheric body" is Steiner's early, theosophical term for the subtle body of supra-physical forces that sustains life. Later he would also refer to it variously as "the life body," or the "formative forces body," or (echoing Spinoza's distinction between *natura naturans* and *natura naturata*) the realm of "living working" as opposed to the physical realm of "finished work." The etheric body is precipitated out of a vast cosmic ether. It is known through Imaginaton, and first reveals itself to strengthened thinking as supra-sensible pictures. The etheric body consists of centrifugal forces, expresses itself in all aqueous processes, and flows in great currents through the cosmos. The etheric is a "time body"; here time becomes space. It is a unity that is always there as a temporal totality, right up to the present moment. Theodor Schwenk's *Sensitive Chaos: The Creation of Flowing Forms in*

Water and Air (London: Rudolf Steiner Press, 1996) is a scientifically compelling and aesthetically beautiful exploration of these forces.

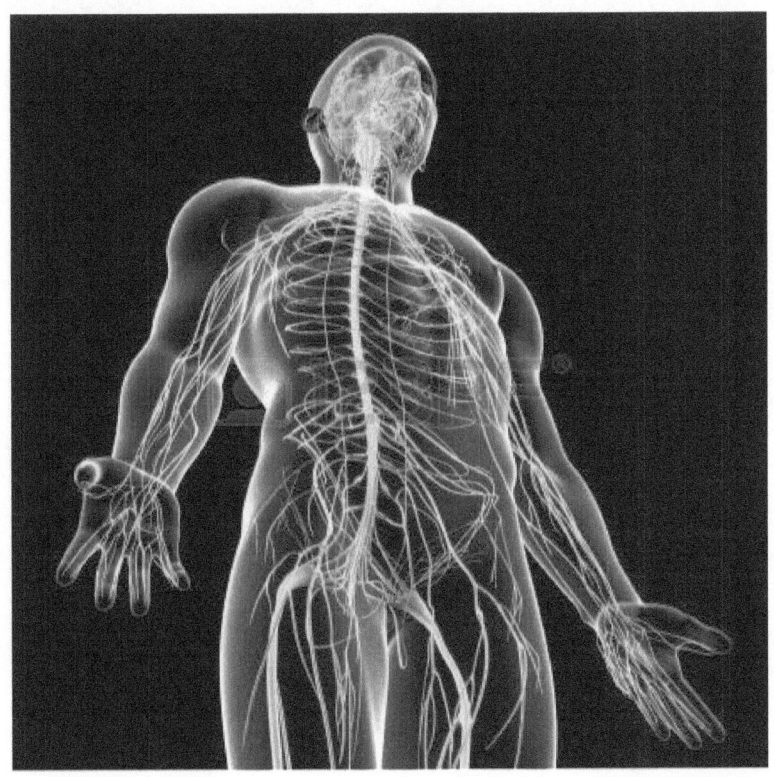

"Astral body" is Steiner's early, theosophical term for the subtle body that corresponds generally to "soul" or "psyche." Like Freud and Jung, he sees it as internally differentiated and gradually transformed by the activity of the higher faculty of the "I" or "ego." The astral body reveals itself to Inspiration, and emerges in a sense from behind Imagination. The traditional concept of the Music of the Spheres is an experience of macrocosmic astrality. It consists of centripetal forces, and expresses itself in breathing and in the airy element generally. It also expresses itself as the human nervous system. The astral body has remained behind in time, and casts its beams forward into the present incarnation; it remains in the spiritual world before conception and birth

The *locus classicus* for both of these bodies among Steiner's introductory works is the uncharacteristically schematic and static description in his early book

Theosophy (1904; many English editions are available, including now a very inexpensive Kindle Edition from Amazon). A much more dynamic (but also much more difficult) account is to be found in the middle four lectures of Rudolf Steiner, *A Psychology of Body, Soul, & Spirit* (New York: SteinerBooks, 1999), which includes a valuable introduction by Robert Sardello. See also Lecture 5 (February 2, 1924) of the cycle GA 234; *Anthroposophy: An Introduction*, trans. and intro. Owen Barfield (London: Anthroposophical Publishing Co., 1961).

Appendix 3

Eurythmy

Eurythmy was an entirely new art of movement that Steiner invented. It was an important episode in the history of dance that has been unjustly neglected. Eurythmy was the continuation of an aesthetic revolution that began not in Europe, but in America. The original impulses leading to the "new dance" were deeply spiritual. It's not the "modern dance" of Wigman, Graham, and Humphreys, but rather eurythmy that is the true heir of the "new dance" – bringing together the spiritual science of Loie Fuller, the balance between Apollonian and Dionysian of Isadora Duncan, and the Oriental Spirituality of Ruth St. Denis. See GA 277a; *Eurythmy: Its Birth and Development*, trans. Alan Stott (Anastasi, 2002); CW 277b; *Introductions to Eurythmy: An Extension of Goethe's Morphological Thinking Within the Realm of Human Movement*, ed. and trans. Frederick Amrine, intro. Gordon Miller ([Amazon:] Keryx, 2017); CW 277c; *The Early History of Eurythmy: Rehearsals and Performances of Rudolf Steiner's* Mystery Dramas*, of the Oberufer Christmas Plays and of Goethe's* Faust*; Addresses, Notes, Programs and Chronologies*, ed., trans., intro. and comm. Frederick Amrine (Great Barrington, NY: SteinerBooks, 2014); CW 278; originally translated under the title *Eurythmy as Visible Music* (London: Rudolf Steiner Press, 1977) but more recently by Alan Stott under the title *Eurythmy as Visible Singing*, 2 vols., 4th edn. (Stourbridge: The Anderida Music Trust, 2013); and CW 279; *Eurythmy as Visible Speech*, translated by Vera and Judy Compton-Burnett (London: Rudolf Steiner Press, 1984).

Appendix 4

Karl Marx

Karl Marx (1818-1883) was a revolutionary socialist and economist. His two most important works by far are his pamphlet *The Communist Manifesto* (1848) and his great theoretical treatise in three volumes, *Capital* (1867, 1885, and 1894). These writings and others, co-authored with Friedrich Engels (1820-1895) became the foundation of the socialist theory known as Marxism.

Marx began as a student of Hegel, one of the so-called "Young Hegelians," but he soon rebelled and developed an antithetical philosophy, dialectical but strongly materialist, that would, as he wrote, "set Hegel on his feet." He spent many years writing a kind of high-level, polemical journalism. A very important precursor to his later, fully developed philosophy is to be found in the *Economic and Philosophical Manuscripts of 1844*.

The revolutions of 1848 radicalized Marx further, prompting him to call for a worldwide revolution of the proletariat in *The Communist Manifesto*. According to his "historical dialectic," the bourgeoisie has shown the seeds of its own dissolution in creating, among other things, a worldwide network of communications and commerce. The proletariat will use this network to eradicate the bourgeoisie and create a more egalitarian society based on expropriation of property and capital by the state. In *Capital*, Marx delivers a withering critique of capitalist social relations at a high theoretical level.

Ironically, Marx enjoyed a very happy, bourgeois family life in England, supported financially by Engels, who was the heir to an industrial fortune. Needless to say, Marx is one of the most influential thinkers of modern times, and his analysis of the excesses of unbridled capitalism still stands. Social experiments carried out in his name, especially Stalinism and Maoism, were highly problematical.

Appendix 5

Eduard von Hartmann

Eduard von Hartmann (1842-1906) was a Prussian philosopher and a theologian. Born in Berlin, he entered the Army as an officer in 1860, but a knee injury in 1861 led to his forced retirement. After spending two years on painting and musical composition, he settled on philosophy as a second career. He obtained a PhD in philosophy from the University of Rostock in 1867, and returned to Berlin. Due to his knee injury, he was an invalid for the rest of his life, often composing in bed, and suffering constant pain. He died and is buried in Berlin.

His first and most important book was *The Philosophy of the Unconscious* (1869, with many editions thereafter). It stood out because of its many concrete examples, and for its vigorous and lucid style. He sought to reconcile rationalism and irrationalism by featuring the central role of the unconscious mind. The unconscious is akin to Schopenhauer's idea of the will, and a self-consciousness conceived along Hegelian lines arises from it. Ideas define the "what" of the world, but will determines its "that." As in Schopenhauer, the idea can eventually emancipate itself from bondage to the will and its torments. This differentiation between unconscious will and the ideas leads to three orders of being which must be distinguished: 1) the metaphysical order of the unconscious; 2) the objective phenomena of nature; and 3) the subjective and ideal order of consciousness.

Above all, *The Philosophy of the Unconscious* was notable for its extreme pessimism. Although this is not the worst of all possible worlds, it would be better for this world not to exist, and indeed the world is moving toward nonexistence.

The book is today chiefly of historical value: it had a large influence in the years immediately following its first appearance, but its influence fell off sharply thereafter. In Steiner's mind, von Hartmann was the greatest living philosopher; he dedicated his doctoral dissertation, *Truth and Knowledge*, to him. (GA 3).

Other important works include *Das Unbewusste vom Standpunkte der Physiologie und Descendenztheorie [The Unconscious from the Standpoint of Physiology and the Theory of Descent]* (1872); *Wahrheit und Irrthum im Darwinismus: eine kritische Darstellung der organischen Entwicklungstheorie [Truth and Error in Darwinism: A Critical Depiction of the Organic Theory of Descent]* (1875); *Kritische Grundlegung des transcendentalen Realismus [Critical Foundation of Transcendental Realism]* (1875); *Die Religion des Geistes [The Religion of the Spirit]* (1882); *Philosophie des Schönen [The Philosophy of the Beautiful]* (1882); *Das Grundproblem der Erkenntnistheorie [The Fundamental Problem of Epistemology]* (1889); *Geschichte der Metaphysik [History of Metaphysics]* (1889-1890); and *Das Problem des Lebens [The Problem of Life]* (1906).

[sources: *Wikipedia* and *Internet Encyclopedia of Philosophy*]

Appendix 6
Ahriman and Lucifer

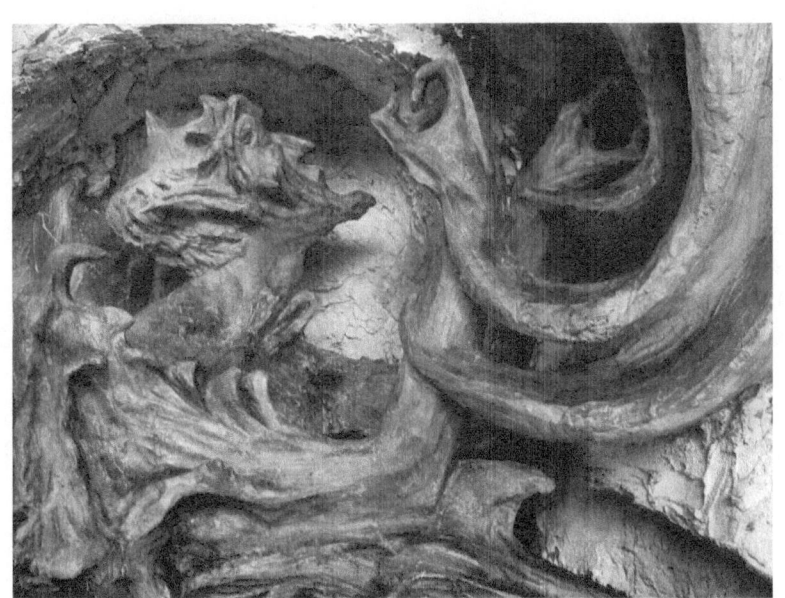

Ahriman

Lucifer

Rudolf Steiner spoke often about the dual nature of evil, ascribing its source to supersensible beings he calls Lucifer and Ahriman. Lucifer might be termed the "red devil," who tempts humans to sin on the side of *superbia*: pride, anger, egotism, erotic passions, etc. "Ahriman" is a traditional name for a black demon, beginning with the Zoroastrian figure Angra Mainyu, opponent of the Sun God Ahura Mazda. Ahriman's temptations are those of *acedia*: laziness, greed, and denial of the Spirit generally.

Lucifer incarnated in the third millennium BCE, in the distant East. Ahriman will incarnate in the 21st century, in the West. Lucifer wants us to live in the past, while Ahriman cuts of off from the past. Lucifer wants us to flee the earth; Ahriman wants to bind us to the earth. Lucifer is responsible for the glories of pagan culture, which however provided no moral impulses. Ahriman seeks to subvert culture by promoting materialism, utility, nationalism, and literalism. He wants to reduce the freedom of the spiritual-cultural sphere to politics and economics. He wants to reduce all qualities to quantities.

Steiner argues that the assaults of these beings are providential: only by overcoming their resistance and holding them in proper balance can humanity become inwardly strong enough to develop genuine freedom, knowledge and love. Both figures are represented in Steiner's monumental sculpture "The Group" (from which these photos have been taken). In "The Group," Christ as the Representative of Humanity shows us how to hold the balance: he does not vanquish Lucifer and Ahriman, but he keeps each in their place, restricting their activity.

For an excellent discussion of Steiner's ideas as applied to Goethe's *Faust*, see Alan P. Cottrell, *Goethe's View of Evil and the Search for a New Image of Man in our Time* (Edinburgh: Floris Press, 1982). Mephistopheles in Goethe's *Faust* exhibits traits of both beings by turns, and Steiner was critical of Goethe for having conflated them. Lucifer and Ahriman also appear as characters in Steiner's own expressionist *Mystery Dramas* (1910-1913): see GA 14, Rudolf Steiner, *Four Mystery Dramas*, trans. Ruth and Hans Pusch (Great Barrington, MA: SteinerBooks, 2007).